**DATE DUE**

| JUN 0 5 2001 | | |
|---|---|---|
| | | |
| | | |
| | | |
| | | |
| | | |
| | | |
| | | |
| | | |
| | | |
| | | |
| | | |
| | | |
| | | |
| | | |
| | | |
| | | |
| | | |

GAYLORD        #3523PI        Printed in USA

# CHOOSING A QUALITY CONTROL SYSTEM

# CHOOSING A QUALITY CONTROL SYSTEM

## Merton R. Hubbard

TECHNOMIC
PUBLISHING CO., INC.

LANCASTER · BASEL

# Choosing a Quality Control System
aTECHNOMIC ᵏpublication

Technomic Publishing Company, Inc.
851 New Holland Avenue, Box 3535
Lancaster, Pennsylvania 17604 U.S.A.

Printed in the United States of America
10  9  8  7  6  5  4  3  2  1

Main entry under title:
   Choosing a Quality Control System

A Technomic Publishing Company book
Bibliography: p.
Includes index p. 205

Library of Congress Catalog Card No. 98-87202
ISBN No. 1-56676-687-7

PRODUCT and service organizations alike agree that a quality control system of some type is desirable, if not absolutely necessary. There are over 24 quality control systems recommended for the control and improvement of quality and process; there are over 30 techniques and buzzwords suggested for implementing them; and to assist in learning about these systems and techniques, there are well over 200 courses, seminars, programs, and conferences available. This book discusses the pros and cons of these many alternatives and suggests how an effective system can be assembled or reconstructed by selecting and combining some basic engineering methods, some nonstatistical methods based on team efforts, and seven statistical tools, with assistance from the power of the computer. This logical system can be expanded and modified as required—later.

Having had an opportunity to observe the different requirements of different company cultures, the author found that there is no one best way to construct or modify a quality system plan that "fits all sizes." This obvious failing has not been made apparent in many highly structured courses to seminar leaders with "ivory tower" backgrounds. And there are CEOs who recognize the need for process control but have neither the background, interest, nor time to devote to it. This book presents the needs, the goals, the cautions, and suggested procedures in simplified terms, which should clarify the four areas to be considered in modifying or constructing an effective system for a company, with some of the precautions (people problems, training, and rewards) highlighted.

The food industry is emphasized, but the suggestions offered should be applicable to hardware manufacturers, process industries, and service organizations. Actual case histories are presented to illustrate how basic nonstatistical quality and process techniques have been used with varying degrees of success. The importance of these components of control and improvement aspects of the system are stressed. Pitfalls of blind trust in computer-assisted procedures are also illustrated.

The single theme is guidance in selecting and assembling the component parts for an effective quality and process system. The major thrust is at upper management, who would benefit the most by having assistance in selecting the right options from a veritable ocean of offerings. Process engineers, store managers, and production personnel would gain a whole new perspective on the basic operations of an effective quality/process control and improvement system. This book should provide guidance in their work in these areas. It should also provide established quality control managers and supervisors with a valuable reference for training quality technicians. It would be of perhaps even greater value to the novice in quality control, since it separates the essentials from the "buzzword of the day."

# CHAPTER 1

# Introduction

CHANCES are the Stone Age ancestor found that, if his hunting and defense weapon consisted of a club-shaped rock fastened to a length of tree branch, he could survive the attacks of small animals by clubbing them on the head. We can also guess that he eventually discovered that, if the edge of the club were chipped off to form a point, it would penetrate the hide of much larger animals, thus significantly increasing the possibility of prolonging his life in the event of an attack. If he were unlucky enough to break this club head with the sharp point, or if he were to lose it, we can be reasonably sure that, when he decided to make a replacement, he would attempt to chip one edge of a rock exactly like the one that worked so well before. This principle of survival is probably one of the earliest examples of quality control: "in order to survive, make it like the last one."

Through the centuries, there were guilds, trades, and craftsmen attempting to teach quality control by the same method: make it like the last one. It was not until Walter A. Shewhart of the Bell Telephone Laboratories sketched a control chart in a 1924 memorandum that the use of statistics was shown to provide a method of quantifying quality. This breakthrough assisted the worker in measuring how similar his product was to the last one he made, using specific numbers for comparison. It clarified why it was possible to make it somewhat like the last one, but not exactly. It applied useful numbers to variability. It led to the concepts of meaningful standards and specifications as well as the many other principles of quality control in use today.

## CHOICES AND OBSTACLES

Over the years, "making it like the last one" has continued to be an accepted base for a quality control program. To this base has been added "make it better, cheaper, and quicker than the last one" "better

1

than our competitor's" and "provide better, cheaper, and quicker services to our customers."

The techniques for accomplishing these added quality requirements are far more complicated than the basic statistical quality control procedures (see Figure 1.1). They vary widely, are occasionally very complicated, sometimes highly theoretical (and impractical), costly to implement, and not equally applicable to all goals, services, products, or companies. They should be used with caution. Some of these techniques are very effective—some are not. Subsequent chapters will discuss this further.

Dozens of excellent quality control techniques have been offered as "The best solution for your quality control," but very few can be successfully applied universally. Worse still, blindly combining two or more diverse systems may result in uncertainty, or even disaster. There are hundreds of books explaining the advantages of one or more systems of quality control. There are hundreds of computer programs containing combinations of approximately 90 quality control-related routines. There are quality control methods that are nonstatistical in nature. This overwhelming abundance of choices can be reduced to seven or eight essentials, but not until a number of obstacles can be resolved.

The first step in choosing a quality control system is to recognize the obstacles that exist in *your* industry, in *your* product, in *your* management, in *your* employees, in *your* methods, *your* suppliers, and *your* customers. Assembling the quality system that is most likely to succeed is possible only after these obstacles have been identified.

In the following chapter, the assorted definitions related to "quality" and "quality control" are discussed in some detail. The intent is certainly not to add to the confusion, but to show the wide selection of "quality principles" that are available. It has been suggested that a company should consider embracing one of the quality philosophies in the same manner that individuals choose to adhere to one religion or another: tradition, pressures of the environment, beliefs, and needs. This is not as radical a proposition as it first appears to be; most company decisions are based on similar philosophical selections. Of the many marketing options available, for example, each company will select the one that has worked in the past (tradition), is most likely to be successful in specific markets (pressures of the environment), is readily accepted by the company sales personnel (beliefs), or is demanded by the customer (needs). Each company will likewise choose a human

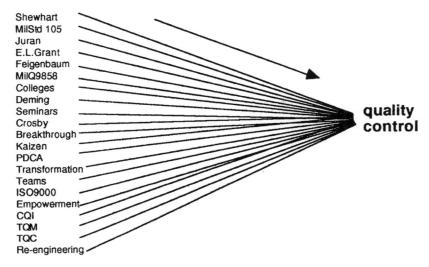

Shewhart
MilStd 105
Juran
E.L.Grant
Feigenbaum
MilQ9858
Colleges
Deming
Seminars
Crosby
Breakthrough
Kaizen
PDCA
Transformation
Teams
ISO9000
Empowerment
CQI
TQM
TQC
Re-engineering

**quality control**

**FIGURE 1.1**  All these roads lead to quality control. Which ones do you take?

resources system, a financing philosophy, or a maintenance plan that works best for its needs.

When it comes to quality control, there is no universal "one best way." A *zero-defects* system might work well for a manufacturer until he discovers that it has locked him into a quality level that is not competitive. On the other hand, a *continuous quality improvement* system might backfire by offering a constantly changing product to customers who expect their reorders to be identical. In some industries, customers might be annoyed by a supplier's constant attempts at *"quality partnering."* Adopting a *total quality management* system, where everybody is responsible for quality, might reveal that, in fact, nobody is in charge of quality. In short, a system that works well for one company may not work at all in another.

So what should your company do to select the best quality control system for its unique requirements? For a starter, let us suggest that the phrase *"select the best system "* be replaced by *"assemble the best system."* This is not a trivial play on words; it is a serious concept that emphasizes the need to develop a system, step by step, that satisfies the specific needs of your organization. The company president may be disappointed that his desire to install a new effective quality control system in two months cannot be achieved if it has to be assembled a piece at a time. The inexperienced quality manager might have his enthusiasm shaken if he returns from a seminar labeled something like

"Total Quality Improvement" or "Reengineered Quality Program" only to find that significant portions of what he has just learned are not applicable to the structure or the philosophy of his company. The machine operator or the fruit juice press employee or the salesperson who completes an in-house training program on a quality-related subject, such as total quality control improvement teams, may come away from the course with the conviction that only a small part of the instruction is applicable in his organization. He may later wonder if any of the quality courses attended are really useful.

At the outset, this might appear to be a rather dismal picture, but it need not be. If the overall goals of a quality control program are clearly understood, a program can be assembled that meets all the requirements unusual or specific to a company. Useful training programs for line personnel can then be selected logically. Corporate programs presented to management can be better understood and accepted.

As a guide to assembling a successful quality program, consider these five basic components and how they can be built into your company:

1. Education and training (quality control principles and techniques)
2. Continuous implementation techniques
3. Engineering (coupled with intuition and experience)
4. Nonstatistical techniques (teams, diagramming, and brainstorming)
5. Statistics (and use of the computer)

The next few chapters will explore each of these components in detail, showing goals, obstacles, and successful examples. Those many popular structured quality control programs offered that do not include all five of these fundamentals may not be doomed to total failure but neither are they likely to come close to total success.

There are countless examples of the wide range of interpretations of quality-related principles found in industry. A mid-1990s *Wall Street Journal* article entitled "Price of Progress," by Al Ehrbar, referred to a few of the then-current buzzwords in the quality area: reengineering, downsizing, total quality management, and statistical process control. With a touch of humor, he dismissed reengineering as a technique that gives firms new efficiency and workers the pink slip. The article describes how a team of *Wall Street Journal* writers spent a year visiting many different types of companies of all sizes in many industries to find out about "downsizing" and quality efforts (among other things). In talking to all of these companies, they asked how important "quality" was. The answer was to forget about quality; forget about TQM—that's a thing of the past. If top management is approached with "we're here to improve quality," it will fall on deaf ears; they are not interested in that. On the other hand, if they are told that "we're here to improve cycle times" or "we're here to show you how you can do more with less," they will be anxious to discuss the ideas.

According to some of the company management personnel interviewed, those who were making major improvements in profits and productivity were those who had divided the organization into work teams and had used "time compression management," "high performance workplace," and other undefined phrases. Some felt that translating the technical knowledge of Statistical Process Control (SPC) into a management tool, not a technical tool, and being "proactive in relations with management" could produce favorable results in quality, productivity, and cost savings. These generalities were suggested with some conviction, *but management was either unwilling or unable to explain the details.* In a few companies, however, management was eager to discuss some of the successes.

One company, faced with a series of failures in modifying their product line, set up a study team, plotted a flow chart, dismantled the entire quality control department, turned them into "coaches," and empowered the workers to do "what they knew best" without interference by management. The results were phenomenal: increase in productivity and major cost reductions.

Another company was asked how they were using the statistical approach to quality control in their office. They replied that the greatest advance using statistics was when they Pareto-charted phone calls in the accounts payable department (in response to increasing supplier complaints directly to upper management regarding late payments). The delay was found to be due to the system requiring the accounts payable department to match 13 data sets of order information with a related invoice before cutting the check. The system was changed: when a shipment was received from a supplier, the receiving dock checked the material with the purchase order, and, if it agreed, the accounts payable department was immediately notified. They, in turn, cut the check within 24 hours. If the order received was incomplete, it was shipped back to the supplier without question. (They only had to do this once or twice before the supplier got the message.) Net result: 15% fewer people in accounts payable, no invoices to process, and payment to vendors now made within 24 to 48 hours.

The *Juran News*, published by the Juran Institute, in the fall of 1994 expressed a similar opinion on the variety of definitions applied to "reengineering." They felt that the word seemed logical and that the implied meaning was appealing but that business literature was vague about its concepts and benefits. The three different approaches might be summarized as

1. Define how work proceeds through key business processes.
2. Start anew with a different look at the way business is conducted by your company.
3. Set up entirely new information systems.

The Juran Institute offers courses in this subject and combines all of the above approaches but defines them more precisely:

1. Business process reengineering
2. Business reengineering
3. Information technology reengineering

It would be a waste of time to agonize over the growing list of quality acronyms, looking for the only one that would work for a given company. It would make more sense to select those portions of systems that are likely to function well with the existing management, work force, product, and customers (see Figure 1.2).

Even after deciding on the systems or portions of systems to be used, the many options for techniques to utilize in structuring the quality system present another series of obstacles.

The late Dr. W. Edwards Deming, considered by most quality control professionals to be a leading authority in the field, recognized several major obstacles to higher productivity and market position. In his book, *Out of the Crisis* (1982), he explains how most of these obstacles are created by industry's ignorance of statistical quality control

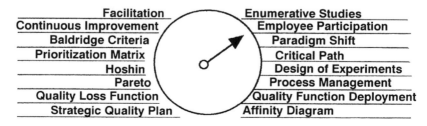

**FIGURE 1.2** Quality techniques—spin the dial to select.

principles or their expectations of the impossible. Following is a very brief summary of this Deming obstacle list:

1. Uninformed management believes that a few days with a quality consultant will produce a complete operating quality system.

2. It is mistakenly supposed that automation, gadgets, and new machinery will transform industry.

3. Some think that blindly copying successful quality control programs from another company will solve their problems.

4. Others think that their company's problems are unique and that principles of statistical quality control will not work for them.

5. Some managers fail to verify the competency of statistical quality control instructors, thus exposing their employees to inadequate quality training.

6. Under some conditions, the use of acceptance sampling tables may guarantee that some customers will receive a defective product.

7. Management that fails to understand control charts and statistical thinking tends to let the quality control department unilaterally handle all quality problems.

8. It has long been thought (incorrectly) that quality troubles rest with the work force, overlooking the need for process improvements.

9. Poorly conceived quality-related programs that subsequently fail produce initial dissatisfaction, followed by frustration and despair.

10. Installation of a complete quality control program is doomed to failure. Such a program must be developed slowly over the years to be effective.

11. An inadequately designed computer system will provide little more than a collection of useless raw data.

12. The supposition that it is necessary only to meet specifications is an open invitation to competitors to steal customers.

13. The concept of zero defects does not correspond to the realities of this world. If forced, it will drive losses and costs to the maximum.

14. Inadequate testing of prototypes that fail to match realistic production conditions usually results in a disaster.

15. A quality control consultant is expected to know all about the business, but this is not necessary. Improvement can come from other kinds of knowledge and nonrelated experience.

Obstacle 10 needs a few words of explanation. The key word here is "complete." Deming certainly does not mean to imply that a complete quality control program is undesirable, for that is the ultimate goal. On the other hand, an attempt to design a complete quality control system from top to bottom, to install it throughout the entire company in a single move, and to expect it to perform successfully is unrealistic. In the first place, most quality control problems are interrelated. If you spend more time on each customer's needs, there may be service delays; if you reduce defect production, storage and shipping may be unable to handle the increased volume; if the net content control is improved significantly, subsequent content measurements, evaluations, charting, and reporting may become redundant; if chemical process control is tightened, entire rework and repair systems might become unnecessary; if computerized production feedback and control is installed, existing control systems might become obsolete, and new untried techniques might be required.

All of these cause-and-effect systems might be predicted, but an attempt to install them without actually seeing the effects over time can easily upset otherwise smoothly operating processes. Perhaps this is what Deming means when he suggests that simultaneous installation of a complete system is doomed to failure. From a practical standpoint, little is to be lost by installing quality control systems one phase at a time. Each step may show improvements and generate enthusiasm—contrasted with the possibility of despair from a complex system that creates new problems.

Obstacle 13 (zero defects) is subject to debate. As a philosophical goal, zero defects is perhaps an admirable principle. As a practical matter, Deming points out that it does not belong to this world. As a motivational tool, it can be partially (and perhaps somewhat dishonestly) achieved by defining "defect" in broad and forgiving terms. For example, if a defective irrigation system drip emitter is defined as one that fails to drip at 15 pounds pressure, sooner or later the laws of statistics will mercilessly combine to produce a defective emitter. If the definition of a defect omits the "at 15 pounds pressure" requirement, it is entirely possible that a "defect" will never be produced. Yet, there still is some value to be achieved by the supervisor's warm-up talk: "let's see if we can get through the whole week without a single defect" or: "let's try to have an entire day with nothing but satisfied customers."

## Failure to Plan Ahead

In addition to the above obstacles found in Deming's book, in earlier lectures he referred to others, which are of interest. For example, over 10 years before job hopping became a common problem for management, he foresaw the need to plan ahead for improved processes with existing equipment and development of new products *with guaranteed tenure of engineers and scientists.* Even if it were possible to guarantee the tenure of the key technical people, two other obstacles promptly appear: tenure of the management and tenure of the marketplace. Should the management change, then in all likelihood, the plans for future products and processes are likely to change with them; should the marketplace change, the need for the proposed new products may disappear. Perhaps Dr. Deming's use of the word "guarantee" is optimistic, but there is no question that planning ahead is mandatory for success.

### Acceptance of Defective Material and Poor Workmanship as a Way of Life

Deming seemed to despair of those who used sampling plans for lot acceptance, claiming that those misguided companies were willing to accept defective raw materials and to produce defective goods as an unavoidable way of life. On the other hand, he was willing to condone the temporary use of these tables provided they led to the adoption of control charts to improve quality of raw material and finished goods. Otherwise, he considered Dodge–Romig sampling plans or the Military Standard Sampling Plans as another obstacle to consistent quality production.

### Detachment of Top Management from Problems of Production: "Our Troubles Lie Entirely in the Work Force"

Two closely related obstacles are management's disinterest in the problems of production, and their belief that all production problems are the responsibility of the work force. However, it is illogical to expect the work force to perform quality miracles with a system presented to them by management. The worker who starts off in the shipping department, moves up to the process line, and becomes production supervisor and finally production manager is fully aware of his responsibility to the line worker. Unfortunately, this is not the route most managers follow. As a result, they are unfamiliar with and perhaps disinterested in the processes.

## Lack of Knowledge About Statistical Quality Technology

While teaching statistical quality control to classes at community colleges, universities, and in company classrooms, instructors have frequently wondered how the students, upon return to their organizations, would transmit the information to their management. How does a student introduce newly acquired quality control information in a frequently hostile environment without jeopardizing his own position?

When is the best time to institute changes learned? One answer is in the familiar quote "the best labor-saving device is tomorrow." But another quote says "he who hesitates is lunch." One of the objectives of this book is to consider alternative ways to find the optimum time to introduce new techniques with minimum distress to both process and personnel. Much of the information has been gained through those who have tried various ways of gaining acceptance of changes in quality and process control methods.

## Total Dependence on Final Inspection and Failure to Use Data from Inspection

Managers who have not been trained in statistical quality technology are likely to create their own obstacle to successful operations by depending on final inspection as the principal measure of product quality. When this antiquated method of operation is used, unsatisfactory results are frequently followed by a combination of slogans and threats, neither of which can be expected to do much good. For those who depend on inspection instead of statistical quality control, inspection tightens during periods of low production, thus increasing reject costs; during rush periods when shipping the product as fast as possible is the order of the day, inspection becomes lax, and anything goes—with the probable loss of customers. Without adequate training, an inspector may produce inconsistent data; without the use of statistical analysis, it may not be realized that the inspectors do not agree with each other in their findings. The use of final inspection systems is not without some merit, provided some type of statistical analysis is applied to the data collected. Unfortunately, this is not often the case, and substandard product continues to be produced.

## The Unmanned Computer

During the period of these Deming lectures, computers were beginning to find their way into the quality control departments to reduce the obstacle of voluminous data collection. Many proved to be effective labor savers; some provided valuable analytical data not previously obtainable; others were useless. Storing masses of raw data in a computer is probably a waste of time. Calculating averages, deviations, defects, and defectives from the data can be interesting. However, to be useful, the data and calculations should be accompanied by control charts, distributions, and other pictorial representations as well as notes explaining the highlights of each chart. Without careful planning for the use of computers, they may present another obstacle to quality control, rather than a solution.

This compilation of Dr. Deming's obstacles is formidable. However, accompanying each obstacle, he has included an inferred solution. For example, in the last obstacle listed above, he suggests that the difficulty with misused computers can be overcome by careful planning before their installation. Rather than attempting to expand on Deming's list of problems, this book will concentrate on quality control obstacles and solutions in five specific areas:

- training
- implementation
- engineering solutions
- nonstatistical solutions
- statistical solutions

These five have been carefully selected because they are common to all of the currently accepted systems of quality control. In addition to exploring the causes and resulting difficulties of these examples, a number of suggestions for improvements in quality control are offered.

## COMPANY QUALITY GOALS

Over the years there has been continuous debate over whether a company's major goal is to provide a product or service, to satisfy the needs of the proprietor, to make a profit for its owners, to offer a source of employment to a community, to generate a flow of tax money to the state for the benefit of its citizens, or all of the above. Regardless of the outcome of this debate, all companies seem to agree on one principle: in order to survive, a company must be managed in such a way that its profits exceed its costs of being in business.

In a manufacturing company, the process costs are readily identified:

| Labor | Materials | Setup | Down time |
| Scrap | Rework | Overtime | Maintenance |
| Warranty | Energy | "Overhead" | Distribution |

and others. Overhead is in quotation marks because it may not be so easily measured as the other items; but, still, it is readily identified.

When it comes to quality costs, there has been general agreement that there are only four sources:

- prevention
- appraisal
- internal failure
- external failure

Based on the premise that quality costs come only in these four categories, many excellent books have been written on the subject, showing the relationships among the factors and how to shift or reduce costs.

Looking at the above simplification of the sources of costs of process and quality, one has an excellent opportunity to focus on likely places to *improve* process and quality and to *reduce costs* of process and quality. Over the years, management's quality control emphasis has narrowed to just these facets of process and quality: improvement and cost reduction. Quality *control* has not been entirely overlooked, but it has been de-emphasized in favor of the concepts of improvement and cost reduction. Deming and others have long maintained that the principles of quality control lead to improvement and cost reduction and that perhaps the emphasis should remain with *control.*

In the next chapter, we will discuss the need for a clear statement of the company's quality goals, be they control, improvement, or cost reduction. Some "clear" statements leave much room for interpretation. This one is from a major electronic company.

- "The quality philosophy of XX Co. is to provide products and services of the highest quality and of the greatest possible value to our customers. Quality is defined as a set of attributes that meets or exceeds customers' expectations." In an interview with one of this company's product managers, he was asked for his understanding of corporate quality. His reply: installing systematic processes necessary to manufacture good products at competitive prices. (This is certainly a very liberal interpretation of the XX Co. definition.)

- Another definition: "A quality product or service is one that meets or exceeds customer expectations." (This policy statement is obviously too brief. If taken literally, the company could exceed its customer's expectations by shipping product a day early—possibly before the customer has room to accept it.)
- A consumer product company: "Our customers shall be assured of consistently superior products." (Superior to what?)
- A bulk food product supplier: "The (products) shall be safe, homogeneous, and of consistent quality, adhering to strict quality specifications."
- A retailer: ". . . shall insure that established company acceptable quality levels are maintained." (This may be a major departure from the concept of the customer's quality requirements.)

These few examples should illustrate the difficulty in stating the company's quality goals, but, without a clear understanding of what the company is trying to accomplish, the departments are likely to stray in different directions in an effort to produce what they believe management wishes the quality level to be. On the other hand, once the quality goal is defined, the choice of applicable quality control techniques starts to become manageable.

## SUMMARY

So far, we have looked at a very bleak picture consisting of choices, obstacles, problems, and misunderstandings that make it extremely difficult to find a satisfactory ready-made quality control system:

Too many quality control systems are offered by the experts
Too many techniques are available to achieve quality control
Management goals may be in other directions (profit, sales, or volume)
Too many obstacles to implementing quality control
Inability to proclaim a meaningful quality policy statement

The remaining chapters of this book are devoted to the five major building blocks of an effective quality control system. They are intended to provide guidance in assembling a system by carefully considering the goals, hazards, and solutions in these five areas:

1. Training management and employees in quality control principles that are applicable to the business is essential.

2. Implementation of a quality control system should be accomplished in incremental steps rather than as a detailed complete program. Identify the many obstacles to effective installation and functioning of a system.

3. Recognize that many types of quality control problems are solved through use of engineering principles not necessarily related to conventional quality control techniques. Include engineering solutions as a portion of the program.

4. Intuition and experience play an important role in quality control, and provisions for both individual and team effort should be formalized to encourage continuous use of these nonstatistical and nonengineering tools.

5. Statistical quality control techniques are readily handled by use of computer programs. The rewards from the use of computerized statistical quality control techniques are well worth the time and cost of their installation, but the associated risks of computer misuse are enormous.

# CHAPTER 2

# Obstacles in Quality Control Training

LACK of training in any profession or technical field is a major obstacle to achievement. Paradoxically, the opportunities for training in quality control in themselves have become an obstacle. The unbridled growth of educational offerings has resulted in a strange combination of opportunity and frustration. Each year, the circle of confusion about training has widened.

There are at least five reasons for this:

1. The multitude of definitions of quality, quality control, quality engineering, and quality management
2. An increasing number of new techniques centered around "quality" and around "control"
3. An increasing quantity of techniques associated with related fields: reliability, process control, quality and process improvement, quality business policies and practices, employee participation, management, problem solving, benchmarking, and more
4. A growing population of academic and industry specialists who offer a wide variety of educational programs and subjects related to some aspect of quality control
5. The addition of new industry and governmental quality standards and quality-based awards

In this chapter, we will examine a half dozen obstacles in quality control training and then suggest some solutions for overcoming these obstacles.

## OBSTACLE NO. 1: QUALITY DEFINITIONS

What is "quality"? This is the first obstacle to overcome in quality training—finding the one acceptable definition of *quality*. In an American Society for Quality pamphlet "Production and Quality," Ross Johnson and William Winchell have selected a definition from ANSI/ASQ 1987 Quality Systems Technology, American National Standard A3-1987. It reads: "The totality of features and characteristics of a product or service that bear on its ability to satisfy stated or implied needs." They have then suggested other acceptable definitions: fitness for use, conformance to requirements, and degree of excellence (although they suggest that this means "relative quality" or possibly "grade").

Each year, students in a quality control class have been asked to define "quality." Each year there is a variety of answers:

- conformance to specifications
- agreement with the blueprint
- six-sigma
- meets consumer needs
- high cost
- provides consumer delight
- top of the line
- perfect uniformity
- classy
- within control limits
- continuous improvement
- a system that avoids quick fixes
- data transformation to improve products

An expensive silk scarf is considered by most people to be a "quality" product. A famous artist's oil painting of a landscape might be considered by some to be "quality art," even though there may be imperfections and inaccuracies in the work. Why wouldn't a photograph of the same subject be considered superior quality, since it doesn't show the inaccuracies or imperfections? An intermediate-priced automobile might be considered "a nice car"; the same car with automatic equipment, special paint job, extra chrome, and a higher price might be judged "a quality car." A sterling silver-encased ballpoint pen might be considered better quality than the same mechanism made with a plastic housing, yet they could have identical performance characteristics.

Perhaps it would be more fruitful if "quality" were defined as "a level of appearance" or "a level of performance" or both. In any case, it is very difficult to strive for, or enforce, quality control until all those people involved in the endeavor can agree on its meaning.

Consider this: one ice cream has a higher butterfat than another. The general public would likely label the higher butterfat product as a *quality product* and the lower fat one as *ordinary*, or even *cheap* or *watery* .

But many might judge the latter as an excellent quality low-fat product. Similarly, a top-of-the-line automobile might be rated by the manufacturer as their *quality line*, compared with the similar, lower-priced, less well-equipped model that they might refer to as a *best-buy* or *top-value* automobile. Yet, the top-of-the-line model might have the reliability of a stubborn mule. Does "quality" refer to appearance and special features only? It certainly isn't intended by the manufacturer, but it might well be perceived that way by the consumer. If the stated need is "it should look classy," then the car is "quality." If the implied need is "it should also start reliably and operate effortlessly without excessive maintenance and without consuming huge amounts of gas and oil," then the quality definition by ANSI/ASQ is valid. Do consumers go through this mental exercise, or do they merely assume these implied needs? Are all consumers alike in their evaluation process? This is doubtful.

Service quality may be even more difficult to define. Does the customer prefer to walk in the door, ask for the service, hand over the money, and walk out, or does the customer believe that "quality service" consists of a casual discussion with the provider, followed by a flexible selection process, a discussion of prices, options, and values, and eventual purchase? Perhaps the customer considers "quality service" to be mail order, telephone transactions, or service by computer. How does management define "quality service" when the customers' preferences consist of all three types?

Admittedly, this battle of words is an obstacle. Until there is agreement on the meaning of "quality" by company management, there's little likelihood that actual "quality control" can be attained.

Another obstacle is the lack of general agreement on the functions of the quality control manager, engineer, or technician. It would be a simple matter to determine what training is required if there were a consensus of opinion as to what these quality control functions should be. In some organizations, the quality control manager has a highly technical job, requiring knowledge and skills in complex electronics, mechanical engineering, advanced statistics, molecular biology. In other companies, the "quality control person is responsible for keeping people from making mistakes." The hospital quality manager must be familiar with medical equipment, testing techniques, medical practices, cost control, drug chemistry and applications, interpersonal relationships, and dozens of other aspects of hospital operations.

"Quality control manager" might be used in almost any situation. The "quality engineer" title sounds somewhat more restrictive, and several attempts have been made to compile the work elements and descriptors. As might be expected, there is a wide variety of opinions. The differing viewpoints of four major organizations are discussed below.

## U.S. Army Management Engineering Training Activity Study

An early major study to examine the meaning of the widely used title "quality engineer" was published as "Survey of Quality Engineering Practices" by the U.S. Army Management Engineering Training Activity in 1978. A list of 94 possible activities of a quality engineer was distributed to 31 military contractors and to 42 private industry companies. The five functions of the quality engineer cited most frequently were:

1. To write or develop inspection or test instructions and procedures
2. To specify product technical and functional requirements
3. To plan or design quality systems
4. To specify quality system requirements
5. To contact suppliers regarding quality

This list contrasted strongly with others found in their literature search, which included the following functions most frequently mentioned as *quality engineering*:

1. Quality cost analysis
2. Analysis of quality data
3. Corrective and preventative action
4. Planning or designing quality systems
5. Development of product quality requirements

The study concluded that proficiency in the following subjects was required for quality engineers: quality cost analysis, quality organization and function, quality information systems, regulations and procedures, and basic statistical quality control.

## The American Management Association

The American Management Association in its job description manual of 1980 defined the duties of the quality engineer as follows:

Designs and installs quality control process sampling systems, procedures, and statistical techniques; designs or specifies inspection and testing mechanisms and equipment; analyzes production limitations and standards; recommends revisions of specifications when indicated; formulates or assists in formulating quality control policies and procedures; develops the economics of any quality control program when required.

## The American Society for Quality

The American Society for Quality has provided peer credentialing for the quality profession since 1968. Professional certification by the ASQ is obtained by written examination. Quality control professionals at various levels who have the experience and knowledge (as evidenced by examination) can achieve the following professional certifications:

- quality manager
- quality engineer
- reliability engineer
- quality auditor
- mechanical inspector
- quality technician
- quality engineer-in-training

The ASQ lists the body of knowledge required for a certified quality engineer beginning in 1995:

1. General knowledge, conduct, and ethics (15 exam questions)
2. Quality practices and applications (55 exam questions)
3. Statistical principles and applications (40 exam questions)
4. Product, process, and material control (30 exam questions)
5. Measurement systems (10 exam questions)
6. Safety and reliability (10 exam questions)

The body of knowledge required by ASQ for a certified reliability engineer:

1. Basic principles, concepts, and definitions
2. Management control
3. Prediction, estimation, and apportionment methods
4. Failure mode, effect, and criticality analysis
5. Part selection and derating

6. Reliability design review
7. Maintainability and availability
8. Product safety
9. Human factors in reliability
10. Reliability testing and planning
11. Data collection, analysis, and reporting
12. Mathematical models

As a guide to the type of courses that might be of special value to an organization, one might consider the types of quality specialists as they are classified by the ASQ. Recognizing the different needs and interests of quality professionals in various industries, the ASQ has established over 20 separate divisions. These divisions provide symposia, workshops, and seminars that focus on the needs of their members. Examples of a few of these divisions are:

- automotive
- biomedical
- chemical and process industries
- customer-supplier
- food, drug and cosmetic
- quality management
- service industries

### California State Board of Registration for Professional Engineers

The State of California, starting in 1975, instituted a process for registration of professional quality engineers. The definition of quality engineer that has been incorporated in California's Rule 404(c-c) reads as follows:

Quality Engineering is that specialty branch of professional engineering which requires such education and experience as is necessary to understand and apply the principles of product and service quality evaluation and control in the planning, development and operation of quality control systems, and the application and analysis of testing and inspection procedures; and requires the ability to apply metrology and statistical methods to diagnose and correct improper quality control practices to assure product and service reliability and conformity to prescribed standards. The above definition of quality engineering shall not be construed to permit the practice of civil, electrical, or mechanical engineering.

Some of the studies and activities that the state listed as embraced by quality control, quality assurance, and reliability included

1. The planning, development, and implementation of inspection and testing techniques
2. The analysis and diagnosis of failures to determine the causes of such failures and the recommendation of corrective action and the disposition of nonconforming materials
3. Expertise in the science of metrology
4. The application of advanced statistical procedures for development of quality control systems and the development of experiments
5. Nondestructive testing and the physics involved
6. Feedback of reviews and recommendations for adjustment of product specifications, manufacturing processes, and equipment
7. Reliability, maintainability, and human engineering to assess or control the probability that equipment and personnel will perform successfully under definite conditions for a specified time
8. Determination that a manufacturing process meets the specified requirements, is stable, and is capable of remaining within the required tolerance at a reasonable cost and in conformance with acceptable standards
9. Review of projects and provision of advice relative to inspection and testing that will be required and other quality-related parameters
10. Expertise in parts standardization and parts application

## OBSTACLE NO. 2: DIFFERING COMPANY GOALS

A press release by the Times News Services (February 1994) noted that a study by Hewitt Associates, a consulting firm, reported that, of 858 employers offering educational assistance to their employees, fewer than 7% of their workers were involved with such programs. Could this low response be caused by a failure of the company to direct the employees to a suitable course choice, or could it be that the employees were unsure of the quality goals of the company?

Before signing up for a course with an intriguing title, it would be wise to ensure that the course will contribute to the goals of company quality control. These goals may be expressed or implied, or even absent entirely.

If a company wants to improve product quality and has not yet established quality control perspective or organization, it would be fruitless to send an employee to a course on quality improvement. It is difficult, if not impossible, to improve on a product's quality unless the current quality level is known and controlled. How can one "make something better" without measuring how good it is now? Trying to make improvements without standards against which results can be measured is a waste of time and effort. If there is no employee capable of determining process capability and the process capability index (Cp), the logical first step toward a quality control program would be to train someone in the basics of statistical quality control. Once the company's process, product, or service control limits are known, then, and only then, should an employee be further trained in improvement techniques.

Another company may have an organization for quality control, but management has no clear definition of quality level, as evidenced by vague directives such as

- "I don't ever want to see one of our products returned because it is defective."
- "We must all pitch in and give the customer the best service."
- "From now on, no more scrap; we are going to increase profits."
- "Our doughnuts shall be equal in quality to our competitor's; our profits must climb; and our costs must drop."

Sending the quality engineer off to study benchmarking, quality function deployment, business process analysis, or other quality-related subjects would be of minimum value unless the quality level could be made more definitive.

A third company may set goals to produce fruit juice equal in quality, value, and price to that of the largest competitor. Simultaneously, the goal for the gross profit of each product line shall exceed 25%. With a clearly defined policy such as this, one would expect the company to benefit from courses of instruction that concentrate on process improvement, quality costs, and strict quality control.

Deming and others have stated that 80% of the quality problems are system problems that can be corrected by management only; the remaining 20% are operator problems. Some have used the figures 85% and 15%, but either will do because both are estimates. If these estimates are reasonable, then sending line employees off to quality control training can, at best, contribute solutions to no more than 20% of the quality problems. On the other hand, training management employees presents an opportunity to provide the tools to solve as much as 80% of the quality problems. In fact, there is always the possibility that management can also devise a system that will eliminate some of the operator problems as well. In short, one of the primary goals should be a decision to train those employees involved in one or the other of these two areas of quality.

Deciding on whether the training should be based on long-term or short-term goals can make a major difference in the type of training required. Long-term goal training classes can be successfully conducted at any location. They would likely emphasize basic principles and theory of quality control. By contrast, training for short-term goals would work best if conducted on the site, with instruction aimed at solving specific company problems.

It is essential to review a company's goals after a training program has been under way. One such procedure is to evaluate the training in terms of its effect on reaching the company's quality goals. Has this expensive program accomplished anything worthwhile? Specifically, is it possible to show that quality complaints have been significantly reduced since the program began? Several other areas to examine include scrap and rework reduction, increased production speed, operating cost savings, fewer line stoppages (or service interruptions), or increased sales attributable to improved product quality.

If these original goals for establishing the training program have not been met, perhaps the program was not the right choice. Some courses based on theory and philosophy of quality control never do get down to the actual tools that the line operator understands and uses daily, or perhaps some of the trained personnel have been replaced with untrained workers, due to promotions or transfers. Possibly some of the initial enthusiasm has worn off.

## OBSTACLE NO. 3: BUZZWORDS

Every profession develops its own vocabulary. The musician refers to tempo, timbre, signature, dynamics, pitch, and many other descriptors. Over the years they have hardened to exact meanings, clearly understood by all musicians. Perhaps the field of quality control is still too young to have developed a vocabulary with precise definitions.

In the early 1950s a bright young man (a new upper management hire who had worked for a major competitor) explained to the company president that, in his former company, "quality control" was considered an archaic term that no longer described the corporate goals. He suggested that the name of the department should be modernized to "quality assurance." The president agreed and promptly sent a memo to all departments notifying them of the change. The quality control manager, confused by this new title, telephoned the president of the American Society for Quality for clarification. His advice—with a hint of a chuckle in his voice—was to accept the new name without making any change in either techniques or goals. "Assurance," he said, was just another in a long line of buzzwords that would eventually be replaced by yet another one.

The use of the expression "buzzword" is not intended to be derogatory. There are dozens of initials, acronyms, and catchy combinations of words that have found wide use and general acceptance. Unfortunately, there is not necessarily a common understanding of their meaning, thus contributing another obstacle to successful quality control. Following is a list of some examples, with suggested definitions.

ANOVA  Analysis of variance
               A statistical technique that identifies and quantifies variations in test data

B  Benchmarking
               Same as quality benchmarking

BR  Brainstorming
               A nonthreatening technique for stimulating free flow of suggestions by a team assembled to improve quality or process

CCP  Constant cause process
               A process in which the variations are constant over time

CI  Continuous improvement
               Improvement of product, process, or service through breakthrough or incremental steps, see also TQM

COPQ  Cost of poor quality
               Includes costs of detection, prevention, and failure

CP  Critical path
               A flow chart of activities required to complete a project, by which the longest path (the critical path) of ad-

joining activities can be determined, also known as PERT Chart

DFQ     Design for quality

DOE     Design of experiments

Generally refers to *statistical* design of experiments, intended to cover all combinations of causes with minimum number of replications

DPM     Defects per million

EVOP     Evolutionary operations

A technique for optimizing a process by making small changes and measuring the effects on the process

FMA     Failure mode analysis

A problem-solving procedure for determining symptoms appearing before and after a failure

FTA     Fault tree analysis

A technique for evaluating the possible causes that might lead to the failure of a product, service, or system

HP     Hoshin planning

A strategy for planning long-term goals for a company, with implementation of work plans based on these "vision statements"

IKD     Ishikawa diagram

Creation of a "fishbone" or cause-and-effect diagram to investigate all causes of a problem

JIT     Just-in-time manufacturing

Incoming materials are received just prior to their need, thus avoiding the cost of raw material inventories

MTBF     Mean time between failures

A (defeatist) term that defines the average time a product will work before failing to function

NDT     Nondestructive testing

Evaluation of a product without damaging it

PC     Process control

Variations fall within specified control limits (not related to statistical control)

PCI     Process capability index

A statistical measure of variability

PDCA     Plan/do/check/act

A system for continuous improvement

PERT  Program evaluation and review technique
See critical path

Q  Quality
See discussion

QA  Quality assurance

1. The function of assuring acceptable quality levels
2. The system whose purpose is to assure that the overall quality control job is being done effectively
3. Same as quality control
4. Confidence that all corrective actions are taken
5. A system that provides confidence that a product will fulfill the quality requirements

QB  Quality benchmarking

1. Comparison of a company's performance against that of the best in the industry
2. Identifying the best performance in a process and, using it as a target, improving the process
3. Evaluating processes others use to solve a task
4. A continuous system of measuring process, product, or service of industry leaders
5. Ranking performance relative to another considered better in a given field and identifying areas for improvement

QBD  Quality benchmark deployment
Same as quality benchmarking

QC  Quality control

1. Same as quality assurance
2. The process of maintaining an acceptable quality level
3. The overall system of activities that provides a quality of product or service that meets the needs of the user

QFD  Quality function deployment
A method of translating a customer's functional re-

quirements into company operations to ensure customer satisfaction

QPM    Quality process management

SMC    Statistical manufacturing control

The process that exhibits only random variations as measured statistically

SPC    Statistical process control

1. The application of statistical techniques to control a process
2. Control of a process where only random variations are exhibited

SQC    Statistical quality control

1. Quality control in which statistical techniques are used
2. Process of maintaining an acceptable level of quality using statistical methods
3. Same as statistical process control

SQM    Supplier quality management

SQV    Systematic quality variation

Variations that exhibit a predictable variation (cyclical, recurring, or trend)

TPC    Total process control

A management system of control, including all departments, to ensure customer satisfaction and optimum costs

TPM    Total process management

Maintaining control of product or service from raw materials through delivery to customer

TQC    Total quality control

Same as total process control

TQM    Total quality management

1. Same as total process management
2. Participation of all members of an organization in improving (or controlling) product, process, and service
3. A technique of defining, measuring, and improving every area of an organization

4. A customer-oriented system that emphasizes management/product/process leadership and human resource excellence

(Note: see more detailed discussion of total quality management in Chapter 3)

ZD    Zero defects
A goal-oriented quality philosophy in which mistakes are not tolerated

The above sketchy explanations may be misleading. There are usually long lists of guiding principles that should be included for their successful implementation. For example, total quality management requires a significant cultural change in which the focus is shifted from the existing company policy to satisfying the internal and external customers. This must be accomplished simultaneously in two areas: the tools of the company and the people of the company. Rather than use the term total quality managent, some companies prefer to invent their own title, such as "quality leadership program."

"Value-added analysis" is one of many buzzwords not included in the above list. It is intended to study each step of the process to determine the cost of the step and the dollar value added to the product at that step. A determination is then made to identify which steps produce the greatest value added or which steps are the most costly to establish priorities for process improvement studies. The same definition comes close to the one that describes "quality cost analysis." These techniques, as well as many in the list, have been used for decades under the general title of "industrial engineering." Often the techniques are renamed to provide the enthusiasm that usually accompanies a fresh start.

Some companies have intentionally avoided the use of buzzwords. At the Archer Daniels Midland Company 70th annual shareholders' meeting in 1993, Chairman James R. Randall said, "Restructuring, downsizing, and other buzzwords are common in industry today. I am frequently asked what ADM is doing in these areas. We are not restructuring because we have nothing that needs restructuring. We have always stayed in the one line of business that we know, and have not become a conglomerate. We are not downsizing because the demand for everything we make is growing. Total quality management is something we have been doing routinely for years. We have not implemented extensive programs in reengineering, broad-banding, skill-based pay, or other current business fads, but have relied on our simple system of logic and doing what makes sense in our daily business."

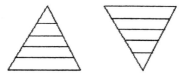

**FIGURE 2.1** Foundation diagram.

In a class with buzzwords, we also find charts, diagrams, and similar artwork. They are intended to clarify a principle and, when constructed with care and used sparingly, are usually very helpful. However, like the buzzword, graphics lose their effectiveness with excessive use.

A commonly used diagram (Figure 2.1) has served to illustrate the increasing (or decreasing) importance of food groups in the diet, top-down or bottom-up management, or the order for adding ingredients to a drink mix.

If these (Figure 2.2) do not become too involved, they may be used to explain sidetracked flow of nonconforming data, products, or systems.

A simple flow diagram (Figure 2.3) is informative in only a general way. It may be used effectively to introduce a principle. However, when developed as an analytical tool, it is more than likely to become extremely complex and confusing to the uninitiated. Fortunately, those preparing a complex flow diagram are most likely the same people who will be interpreting it.

Most companies find that this readily recognized chart form (Figure 2.4) is an excellent way to indoctrinate new employees. For huge organizations, it is generally broken down into assimilable subcharts. In practice, it is probably not a true representation of the reporting road map. Some companies are attempting to operate without such a chart, hoping to function as "one big team." Sooner or later, this leads to confusion.

A simple chart (Figure 2.5) is widely used and clearly shows the major steps and the order in which they are taken—very helpful.

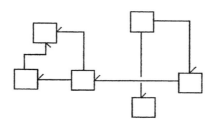

**FIGURE 2.2** If/then diagram.

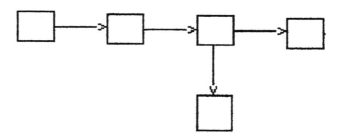

**FIGURE 2.3** Simple flow diagram.

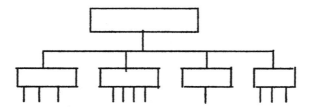

**FIGURE 2.4** Organization chart.

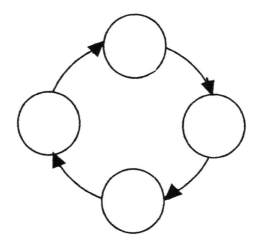

**FIGURE 2.5** Plan/do/check/act chart.

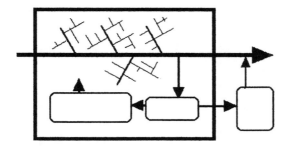

**FIGURE 2.6**  Cause and effect diagram.

Simple diagrams, such as that shown in Figure 2.5, are easily understood and make effective learning tools. Now, a few examples of charts that become obstacles to learning will be illustrated.

Figure 2.6 may be of some use to the individual who prepares it but can be extremely difficult for others to comprehend. The diagram above (with additional descriptive words) was actually used in a quality control class as a teaching aid. The effect on the class was devastating.

Figure 2.7 suggests that the flow of information from department to department is clearly established. One might expect, however, that the student, after studying this chart for a minute or two, would probably conclude, rightly or wrongly, that everybody communicates with everybody else.

A matrix such as Figure 2.8, with 30 entry blocks, is about the limit of complexity that a student should be expected to handle. As the matrix size exceeds this, the ability to comprehend the interactions of the lines and columns begins to fade, and the matrix, instead of a teaching aid, then becomes an obstacle to learning.

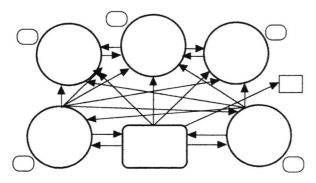

**FIGURE 2.7**  Communications chart.

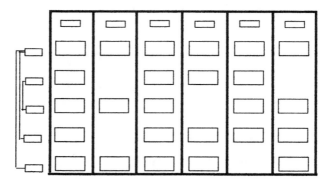

**FIGURE 2.8** Matrix.

There is no limit to the types of charts or the symbols contained within them. One picture is worth a thousand words, but a thousand pictures become a useless exercise. The diagram of the month quickly loses its effectiveness. As a case in point, in one large company there is a room set aside for copy machines. Because of the sizable cost of operating these machines, the company created a graph showing the cost of paper used each month. A visitor to the plant was enthusiastic over this innovation, until he noticed that the last entry on the graph was 8 months old!

Before leaving the discussion of buzzwords and overworked charts, we should mention one more teaching aid that has become so prevalent that it has lost its effectiveness. This is the use of summing the number of items in a list of techniques and proclaiming them as the 12 tools to the road to success, 17 steps to quality improvement, etc. The appearance of yet another list of steps to the promised land is likely to generate student lethargy—if not outright antagonism. The quality gurus may possibly have been the cause of this situation. For example,

- The four steps of the Deming-Shewhart cycle
    1. Plan
    2. Do
    3. Check
    4. Act
- Juran's seven steps (Juran and Gryna, 1970)
    1. Establish quality policies.
    2. Establish quality goals.
    3. Design quality plans.

    4. Assign responsibility.

    5. Provide necessary resources.

    6. Review progress against goals.

    7. Evaluate manager performance.

- Deming's 14 steps (Deming, 1986)

    1. Create constancy of purpose.

    2. Adopt the new philosophy.

    3. Cease dependence on mass inspection.

    4. End awarding business on the basis of price.

    5. Constantly and forever improve the system.

    6. Institute on-the-job training.

    7. Provide leadership.

    8. Drive out fear.

    9. Break down barriers.

    10. Eliminate slogans and targets for zero defects.

    11. Eliminate work standards and management by objectives.

    12. Remove barriers to pride of workmanship.

    13. Institute a vigorous program of education and improvement.

    14. Enlist everyone to accomplish the transformation.

- Feigenbaum's three principles (Armand V. Feigenbaum, 1983)

    1. Make boardroom decisions on quality as strategic goal.

    2. Create a system of quality management and technology.

    3. Establish the continuing quality habit.

- Tom Peters' five principles (Peters and Waterman, 1982)

    1. Offer abiding management commitment.

    2. Institute wholesale empowerment of people.

    3. Involve all functions of the firm.

    4. Create encompassing systems.

    5. Emphasize customer perceptions over technical specifications.

- Crosby's 14 stages (Philip Crosby, 1979)

    1. Management commitment

    2. Quality improvement team

    3. Quality measurement

    4. Cost of quality evaluation

   5. Quality awareness
   6. Corrective action
   7. Establish zero-defects program committee
   8. Supervisor training
   9. Zero Defects Day
   10. Goal setting
   11. Error cause removal
   12. Recognition
   13. Quality councils
   14. Do it all over again
- Deming's seven deadly diseases (Deming, 1986)
   1. Lack of constancy of purpose
   2. Emphasis on short-term profits
   3. Evaluation of performance and annual reviews
   4. Mobility of management—job hopping
   5. Management by use only of visible figures
   6. Excessive medical costs
   7. Excessive costs of liability

One should not be surprised if presentation of yet another list of steps to the golden glory of quality improvement is greeted with "Lord! Not another one!"

## OBSTACLE NO. 4: CONFUSED GROWTH OF EDUCATIONAL OFFERINGS

Starting in 1968, the American Society for Quality sponsored a study entitled, "Establishing Educational Programs in Quality and Reliability." A group of nine highly qualified quality professionals published the initial report in 1969 and revised it every few years. By 1982, the report documented ten universities that offered a Bachelor of Science degree with majors in quality. Some of the course offerings at the Bachelor of Science level were

- Control Aspects of Quality
- Data Processing for Quality Control
- Design of Experiments
- Economics of Quality

- Elements of Quality Assurance
- Gage and Tool Control
- Introduction to Quality Assurance
- Nondestructive Examination and Methods
- Principles of Dimensional Inspection
- Principles of Quality Assurance
- Principles of Reliability Assurance
- Process Control Engineering Technology
- Processing Quality Control Data
- QA Applications
- QC Organization and Management
- Quality Assurance Organization and Management
- Quality Capability Analysis
- Quality Control Manual
- Quality Costs
- Quality Economics and Vendor Quality Control
- Quality Information Equipment Engineering Technology
- Quality Methods and Standards
- Quality Systems
- Sampling Technique Analysis
- Software Quality Assurance
- Special Quality Experiments
- Statistical Quality Control
- Total Quality Control

Surprisingly, the following year, not one of these ten universities offered a B.S. in quality control. A few continued to offer specialized quality courses, but it seems that the limited demand for a B.S. failed to justify maintaining the specialized degree facilities. Since that time, a few universities have again offered this degree, and several have reinstituted courses related to quality.

By contrast, in the same report, 16 community colleges offered Associate in Science degrees in quality control, quality technology, or reliability and have been joined by others since the report. Courses are added or dropped, depending on local industry needs and college budgetary constraints. The core courses offered by most community colleges generally include

- Quality Control Concepts and Techniques
- Statistical Quality Control Theory and Practice
- Quality Control Engineering Principles and Techniques

*Table 2.1.  Degrees with "Quality" or "Reliability" in the Title (%).*

| | Colleges and Universities | | | Community Colleges | |
|---|---|---|---|---|---|
| | Number Surveyed | Bachelor | Master | Doctor | Number Surveyed | Associate |
| 1982 | — | 10 | 0 | 0 | — | 6 |
| 1983 | — | 0 | 0 | 0 | — | — |
| 1992 | 160 | 6 | 11 | 1 | 60 | 21 |
| 1993 | 139 | 7 | 31 | 5 | 46 | 21 |
| 1994 | 206 | 7* | 13** | 5 | 74 | 27 |
| | "quality related" or "quality concentration" | | | | | |
| 1994 | | 14 | 27 | 6 | | 8 |

*One of these is in Canada.
**One of these is in Mexico, and one is in Scotland.

- Reliability Techniques
- Nondestructive Testing
- Metrology (mechanical, electrical)
- Other quality-related subjects, as requested locally

After looking at this brief list, someone is bound to ask, "How about Quality Costs?", "Why isn't Quality Management included?", "Where is Quality Standards?". There is no "one size fits all" curriculum.

Slowly, there has been a reawakening of offerings for degrees in quality control at the university level. The American Society for Quality conducted several surveys among the U.S. colleges. A comparison of earlier and recent surveys is summarized in Table 2.1.

Because of the differences in the populations sampled, it might be useful to convert these data to percentages, as shown in Table 2.2.

At the university level, during these 3 years, there were very few op-

*Table 2.2.  Degrees with "Quality" or "Reliability" in the Title (%).*

| | Colleges and Universities | | | Community Colleges |
|---|---|---|---|---|
| | Bachelor | Master | Doctor | Associate |
| 1992 | 4 | 7 | 1 | 35 |
| 1993 | 5 | 22 | 4 | 46 |
| 1994 | 3 | 6 | 2 | 36 |
| "Quality related" | 7 | 13 | 3 | 11 |

*Table 2.3.*

| Colleges and Universities | | | Community Colleges | | |
|---|---|---|---|---|---|
| Number Surveyed | Offer Quality Courses | Offer Degrees or Quality Minor | Number Surveyed | Offer Quality Courses | Offer Quality Control Certificates |
| 177 | 53 | 27 | 59 | 28 | 22 |

*Source: Quality Progress,* September 1997, pp. 31–65, Vol. 30 No. 8.

portunities to obtain a basic bachelor's degree or a doctorate in the quality field. The master's degrees increased in availability during 1993, but then subsequently returned to the 1992 level.

It appears that the total number of associate degrees has remained fairly level at between one third and one half of the community colleges.

A later survey (1997) by the American Society for Quality changed the format; thus a direct comparison is not possible. The data collected indicate no drastic change in the trends of quality education at colleges and universities or at community colleges (Table 2.3).

It is all but impossible to find a university offering a degree in *quality control.* For some reason, the academic world has seen fit to coin dozens of other descriptions of the discipline, most of which include the word "quality," "reliability," or "assurance." Table 2.4 shows the distribution of quality degree titles that were reported in the 1993 survey.

*Table 2.4. Degrees with "Quality" or "Reliability" in the Title (%).*

| | Colleges and Universities | | | Community Colleges |
|---|---|---|---|---|
| | Bachelor | Master | Doctor | Associate |
| Business Admininistration in Quality Improvement | 0 | 1 | 0 | 0 |
| Business Admininistration in Quality Management | 0 | 1 | 0 | 0 |
| Business Admininistration in Quality Science | 0 | 1 | 0 | 0 |
| Product Assurance Engineering | 0 | 1 | 0 | 0 |
| Quality | 1 | 2 | 0 | 0 |

## Table 2.4 (continued). Degrees with "Quality" or "Reliability" in the Title (%).

|  | Colleges and Universities | | | Community Colleges |
|---|---|---|---|---|
|  | Bachelor | Master | Doctor | Associate |
| Quality and Human Resources | 0 | 1 | 0 | 0 |
| Quality and Reliability | 0 | 0 | 1 | 0 |
| Quality and Reliability Assurance | 0 | 1 | 0 | 0 |
| Quality and Reliability Management | 0 | 1 | 0 | 0 |
| Quality Assurance | 4 | 3 | 1 | 6 |
| Quality Assurance Technician | 0 | 0 | 0 | 2 |
| Quality Assurance Technology | 0 | 0 | 0 | 3 |
| Quality Control Technology | 0 | 0 | 0 | 3 |
| Quality Costs | 0 | 1 | 1 | 0 |
| Quality Engineering | 0 | 4 | 0 | 0 |
| Quality Engineering Technician | 0 | 0 | 0 | 1 |
| Quality Engineering Technology | 0 | 0 | 0 | 1 |
| Quality Improvement | 0 | 1 | 0 | 0 |
| Quality Improvement Specialist | 0 | 0 | 0 | 1 |
| Quality Industrial Engineering | 0 | 0 | 0 | 1 |
| Quality Management and Productivity | 0 | 1 | 0 | 0 |
| Quality Management | 0 | 5 | 1 | 1 |
| Quality Sciences | 1 | 1 | 5 | 0 |
| Quality Systems | 0 | 1 | 0 | 1 |
| Quality Value Leadership | 0 | 1 | 0 | 0 |
| Reliability | 0 | 1 | 0 | 0 |
| Statistics and Quality Improvement | 0 | 0 | 1 | 0 |
| Total Quality Management | 1 | 3 | 0 | 1 |

Sources: Quality Progress, October 1992, pp. 41–60, Vol. 25, No.10; Quality Progress, October 1993, pp. 31–60, Vol. 26, No. 10; and Quality Progress, September 1994, pp. 47–78, Vol. 27, No. 9.

## Seminar and University Quality Control Courses

Let us look at seminars and university courses. Too often have consultants and seminar instructors been harshly described as theorists and academics who haven't the foggiest notion of how things work in the real world. Before sending personnel out to a training program, check out the speakers or instructors. What actual, on-line experience have they had? Are they still working in their field? If not, how long have they been away from it? Have they lost touch? Are they teaching some kind of theoretical concept that is too complex to function in your organization?

Seminar announcements make frequent use of anecdotal examples of successful applications (or failure of competitive seminar applications). More often than not, these may be overblown as generalities, even though they may have been effective only once.

Some of the ideas promoted might be designed to take three to five years to implement, during which time perhaps half of the indoctrinated employees will have been promoted to new jobs or possibly will have left the company altogether. It might be necessary, six months after the start of the new program, to plan on setting up an in-plant tutoring program to bring new employees up to speed, or, in the event of a major management change in the middle of the training program, will it have to be dropped?

Some courses promote a change of culture to introduce a new concept of quality control or process improvement. Suppose that one of the culture changes requires top management to share knowledge concerning corporate financing, market plans, engineering projects for the future, costs of raw materials and supplies, expansion or relocation projects under way, or other operations and directions that are currently unavailable to any department or any individual not directly involved. If the seminar attendee finds that some of this type of information sharing is an integral part of the culture change required to create an improved quality or process improvement, how can he possibly expect to single-handedly turn the corporate philosophies upside down? This is one reason it is important to understand the major principles of the subject matter taught in quality control courses before signing up for them.

## Who Should Attend?

Assuming that the course has been thoroughly researched, and it is decided that someone should spend the required three days in Chicago or Toledo to learn about the techniques, how does the company go about selecting the right candidate to attend? Chances are that most of the time the individual reading about the courses to be offered asks for permission to attend because of a "sincere interest to become more expert in this field so that I may be of greater value to the company"—or some other seemingly compelling statement. Every attempt should be made by the candidate's supervisor to find out if more selfish motives exist. Atlanta might be the very city to where an employee's boyfriend happened to move last month when he was transferred. In another case, a promising young quality laboratory technician asks for permission to attend an intensive two-week course in basic statistical quality control so that he can move up in the quality organization, perhaps becoming quality manager in one of the company's plants some day. The very day he returns from the course, he gives two weeks' notice; he had been offered a position as quality manager in another company. Yes, he knew about the offer when he asked for permission to attend the training program.

Supposing the manager of research and development reads about a quality control course and suggests that Quentin, the quality control manager, attend. Unless Quentin is particularly enthusiastic about the idea, chances are he will say something like, "I'm right in the middle of a new computer program installation that has to be finished by the end of the month. Perhaps I can send my second assistant, Charlie. We don't really need him here next week because line #9 is going to be shut down for maintenance." That is obviously not a valid reason to send Charlie, but, all too frequently, the person who is available is the one that goes—a dumb idea.

Sometimes somebody near the top of the management echelon reads about a college course relating to quality control and decides to go himself, even though his interests and background have very little to do with quality control. If the course requires prior knowledge of statistical analysis, which the manager does not possess, this can be a complete waste of time. Fortunately, there are many occasions when upper management has the good fortune to attend a lecture by one of the accepted gurus of quality control, and he returns to the office with a far better understanding of the concepts his quality control manager has been promoting. On the other hand, if the course is built around some theoretical new buzzword, he might return to present the quality control manager with a fruitless new project on which he is required to spin his wheels.

So, who should attend? Obviously, there is no one best answer to this question. To study new corporate quality philosophies, top management should enroll. Quality theories and techniques courses are usually designed for quality supervisors and frequently may be of value to nonmanufacturing managers: purchasing, accounting, order entry, marketing, personnel, maintenance, and customer service. High-level quality improvement offerings are frequently aimed at quality engineers and design engineers. Basic quality control statistics and charting courses are most useful to supervisors and quality control technicians and owners or managers of very small companies. Studies on team concepts and simple quality control tools should be offered to employees at the order desk, on the line, in the field, and anywhere direct customer contact is part of the job, *provided that their supervisors have also attended the same courses and plan to use team principles.*

### Promotional Course Material

The advertising for university courses and seminars that are centered on quality is designed to inspire an overwhelming desire to attend. The goals, claims, and the testimonials are convincing. Often the course title is promising, glorified, mystical, tempting, or bullying. Although course outlines are sometimes included, the terms may be obscure. Here are some examples selected from brochures describing the hundreds of courses offered:

#### Total Quality Management

Total quality management, the standard for today's participation in global competition, is essential for survival in today's global economy. World class firms of the future will be those that combine product and service quality, flexibility, and effective employee involvement.

#### Testimonials:

"Excellent overview of the spectrum of what TQM can do, and some key tools."
"A complete view of focusing in on TQM."

#### Business Process Analysis

Every company is examining how the work actually gets done versus how it should get done. Who can afford to waste time, resources, and money? Attend this premier conference and reap the rewards of process improvement.

Testimonials:
"One of the best conferences I have attended recently that can be utilized for immediate value to the organization."
"All information presented is relevant, well presented, and can be utilized immediately in my own facility."

## Quality Function Deployment

Quality Function Deployment translates the voice of the customer throughout the various stages of product or service development, delivery, and beyond. Development times are shortened and costs are reduced by establishing priorities and focusing on what is most important to the customer and to the organization's competitive position.

## Implementing Continuous Improvement

Over 20 years we have created a series of business buzzes that possibly surpass anything of their kind this century. We called it *Quality Circles, Culture Change, Total Quality, Customer Focused Quality,* ad infinitum. Yet, at its simplest, it is the implementation of an unrelenting focus on one thing—continuous improvement.

## Best Practices (from over 100 World Class/Total Quality Leaders Around the Globe)

Your team can produce 5 to 10 to 20-fold improvements with less staff, time, and resources. Dramatic reengineered and continuous improvements in quality, speed, service, and cost—regardless of what your organization does.
Testimonials:
"Very good job of helping me benchmark where my company is versus other industries in some highly important measurements."
"This has been the best organized, most useful information-packed seminar I have ever attended."

## Benchmarking High-Performance Work Teams

Strategic approaches for increasing productivity, quality, and customer satisfaction. An exclusive approach to high-performance teams.
Testimonials:
"This conference was of high quality. It crystallized JIT/TQM/simplicity/employee empowerment into a full business strategy."
"The expertise of the presenters and the willingness of all to share information exemplified the spirit and intent of the conference."

Introduction to Process Management
Presenting a set of tools for identifying, analyzing, improving, and managing the processes that impact quality, cycle time, cost, productivity, customer focus, and organization structure.

## Community College Quality Control Courses

Community college courses in quality control generally have fewer mysteries to explain. First, the courses offered are generally less complex than some of the seminars and university presentations, possibly because most of the community college courses are aimed at a different audience. The market for local education generally consists of nearby employees. It requires courses constructed at a practical level taught by experienced instructors in specific quality fields and designed for students from industries predominant in the area. The course outlines are generally concise and clear, with titles that are generally well understood by those in the field. Often the instructor and course content are subject to review by members of a professional quality control organization so that students may be assured of current authoritative information.

In spite of the undeserved stigma of inferiority sometimes associated with community college education, students from all levels of labor and management have benefitted from these courses. Often the presence of a few higher-salaried students in the classroom and the "back-to-school" atmosphere coupled with the predominantly nighttime courses contribute to the attentiveness (and, frequently, enthusiasm) of the students.

An informative 76-page report was published in 1969 entitled *Quality Control Program Study for Community Colleges.* The study that produced this report was conducted by the California Community Colleges, utilizing a mix of technical education specialists and industry quality control professionals. Initially, 107 industries (with an aggregate employment of 13,400 employees engaged in quality control work) responded to a detailed questionnaire and follow-up telephone interview, and in all cases they indicated a continuing need for trained quality control technicians. It was determined that the community colleges could best serve industry by providing courses geared to quality technicians rather than quality or reliability managers or engineers.

The courses selected to meet this requirement were based on seven areas:

- an introduction to the philosophy of quality control
- electrical and mechanical quality measuring techniques
- basic quality control statistics
- specification and drawing interpretation
- applied physics
- technical mathematics
- technical report writing

Over the years, community college curricula have been modified to meet current needs. A few examples of quality control and reliability curricula, leading to an Associate in Science degree, have been selected from a publication of the ASQ Education and Training Institute:

Example 1. Associate in Science Degree, Quality Technology
    Semester 1
        Quality Control Concepts and Techniques
        Statistical Concepts and Techniques
        QC Communications and Human Factors
        English
        Political Science
    Semester 2
        Statistical Applications
        Quality Control Engineering Principles and Techniques
        Electronics Quality Control
        Psychology
        Mathematics
    Semester 3
        Quality Control Engineering and Applications
        Introduction to Nondestructive Tests
        Reliability Objectives
        English
        Math
    Semester 4
        Mechanical Metrology
        Electrical Metrology
        Speech Arts
        Health Education
        Physical Education
        Supervision

Example 2. Associate in Science Degree, Quality Control and Reliability

First two semesters
 Communications
 General Chemistry
 College Algebra
 American Government
 Quality Control Concepts and Techniques
 Physical Education
 Fundamentals of Speech
 Introduction to Applied Physics
 College Trigonometry
 Human Relations in Industry
 Engineering Drawing
Final two semesters
 Quality Control Engineering
 Reliability Objectives
 Technical Writing
 Introduction to Statistical Quality Control
 Quality Control and Reliability Management
 Statistics
 Group Process
 Statics and Strength of Materials
 Statics and Strength of Materials Lab

Example 3. Associate in Applied Science Degree, Quality Assurance Technician

 Two-year evening program
General Education
 Communications Skills
 Reporting Technical Information
 American Institutions
 Economics
 Psychology of Human Relations
Technical Support Studies
 Technical Mathematics
 Technical Science
 Physical Metallurgy
Technical Studies
 Materials and Processes of Industry
 Welding Design and Applications
 Quality Assurance Concepts and Techniques
 Principles of Nondestructive Testing

Inspection Standards
Statistical Concepts
Metrology
Nondestructive Testing Application and Practice
Quality Control Theory and Application
Practical Problems in Quality Assurance

One of the pioneer community colleges offering an A.S. degree in quality control and reliability is located in an area with a high concentration of aerospace industries. Requirements are as follows:

25 units of major courses
  Quality Control Concepts and Techniques
  Quality Control Engineering Concepts and Techniques
  Quality Control Engineering Theory and Applications
  Statistical Quality Control—Theory and Practice
  Reliability Objectives
  Introduction to Nondestructive Tests
  Mechanical (or Electrical) Technology
Six units of major electives selected from
  Specialization in Supervision
    Management Concepts
    Reliability Techniques
    Manufacturing Data Processing
    Configuration Management
  Specialization in Nondestructive Testing
    Radiography
    Radiographic Safety
    Ultrasonic Testing
    Eddy Current
  Specialization in Inspection and Metrology
    Procurement Quality Control
    Electronics Quality Control
    Mechanical Metrology
    Electrical Metrology
Six units of electives
  Engineering
  Data Processing
  Materials Management
  Mathematics
  Metallurgy

Quality Control and Reliability
Purchasing and Supervision
23 units of general education
English
Political Science
Psychology
Data Processing
Health Education
Fine Arts and Humanities
Physical Education

Another community college with students predominantly from the computer industry has enrolled hundreds of A.A. quality assurance degree candidates. Their course requirements have been surprisingly similar to those of the aerospace industry community college.

Major requirements
Human Relations in Business
Concepts of Physics
Introduction to Quality Assurance
Introduction to Statistical Quality Control
Quality Control Applications
Statistical Concepts and Techniques
Total Quality Assurance Concepts and Techniques
Basic Electronic Inspection
Four courses from the following
Reliability Objectives
Quality Control and Reliability Management
Introduction to Nondestructive Testing
Principles of Electronic Testing
Introduction to Quality Data Management
Introduction to Governmental Requirements
Quality Engineering
Advanced Techniques of Nondestructive Testing
Quality Technology
Computer Software Quality Assurance
Reliability Engineering
Quality Analysis
Quality Systems Audit

The St. Louis Community College Center for Business, Industry and

Labor (CBIL) created a series of courses for one of their city's major manufacturers to train all of their hourly employees in basic computer techniques. Additionally, CBIL developed a program in problem solving in special quality areas common to that industry. Other community colleges have offered quality control courses that have been presented at individual company locations, with community college credit being given to those who successfully completed the requirements.

Because community college curricula are tailored to the demands of the industries in the areas they serve, it is vastly simpler to select applicable courses from their catalogs than to agonize over the other countless educational offerings.

## Conferences

For the most part, conferences are designed around some theme usually related to problems or solutions of the day. There is a thin line of distinction between a seminar and a conference as these terms are currently used. A seminar used to be defined as an advanced group of students studying under a recognized leader in some prescribed field, where each student exchanges results of original research through reports and discussions. A conference was more simply defined as a formal interchange of views.

In the professional world, the *exchange* or *interchange* portions of these definitions have faded. For the most part, professional conferences consist of lectures on a variety of subjects related to a central theme, with a minimum amount of discussion in the form of a few questions from the audience. Seminars are more often presented by a single lecturer, with more opportunity for audience participation, depending upon the size of the audience. The classical definitions above both still apply to some college classes; the conference definition is more suited to business meetings within a company.

The obvious advantage of conferences and seminars over college courses is that they are brief, usually a single day, and rarely extend beyond a week. They generally are concentrated on a few subjects, although annual conferences might cover a multitude of subjects, combined in groups of sessions.

As discussed previously, the disadvantage of signing up for seminars and conferences is the problem of evaluating their usefulness, based on a reading of their publicity releases. Most of the conferences are carefully prepared and tested by reliable organizations and can be of considerable value to companies interested in updating their quality control and improvement efforts. A few, on the other hand, are untried programs that are advertised by individuals with the hope of signing up enough attendees to provide a reasonable profit after hotel meeting room bills are paid. If the response is too low, the program is canceled. There are methods to evaluate such advertisements, and they are discussed later.

The American Society for Quality Annual Technical Conference is held in a different city each year. Typically, it consists of three days of committee meetings, three days of technical sessions, a few special tutorials, and several days of commercial and professional exhibits. Approximately nine technical sessions are offered in the mornings and nine more in the afternoon, with one to three presentations in each session. Each session covers a distinct classification of subjects selected to encompass the theme of the year, thus offering "something for everybody."

Other ASQ-related conferences are offered by divisions, regions, and sections. Divisions organize one- to three-day conferences, concentrating narrowly on subjects in their specific field. Regional events are more general and may be considered as miniature national conferences. Those presented by sections are generally restricted to subjects of interest to local industries. Each of these smaller events provides an opportunity to meet quality control professionals with similar interests, thus promoting discussions that may be beneficial to all.

Another national quality related conference is the Quality Expo International Conferences and Trade Show, sponsored annually by a magazine publisher. The structure of this show, with six tracks of sessions, is quite similar to the ASQ Conference, but the emphasis is somewhat more closely allied with the metrology field. A recent program included such sessions as:

- Design and Tolerancing
- Six-Sigma Quality
- Managing Total Quality
- Statistics for Quality
- Coordinate Metrology
- Defect Management Tools
- Conformity Assessment

## *The Transformation of American Industry Program*

A novel modular system was developed in the mid-1980s to build a national community college program for teaching techniques to improve quality and productivity. The project was sponsored mainly by a number of Midwestern organizations, generally related to the automotive industry. It was coordinated by the Jackson Community College in Jackson, Michigan, and was endorsed by the American Association of Community and Junior Colleges.

By borrowing Dr. W. Edwards Deming's phrase, "transformation of industry," this program has acquired the prestige of Deming's teachings and has a distinct advantage over a "quality control" title. The course consists of 12 modules, each of which is devoted to a single aspect of quality control and improvement. The modules are summarized in carefully structured pamphlets of 10 to 50 pages that outline each phase of the subject. Half-hour videotapes accompany each module for classroom use. In addition, an instructor's guide assists in presenting a uniform teaching technique. The module titles listed below suggest an overview of the material covered:

1. The Transformation of American Industry
2. Deming on Quality and Productivity
3. The Transformation Process: Model for Quality/Productivity Improvement
4. Project Selection
5. Project Implementation: Data Gathering and Problem Solving
6. Project Implementation: Data Analysis and Interpretation
7. Project Implementation: Process Control (X–R Charts)
8. Project Implementation: Process Capability
9. Project Implementation: Median and Individuals Charts
10. Project Implementation: Attributes Charts
11. Project Evaluation
12. Continuing Improvement Strategies

The course displays a strong influence of the automotive industry and generously borrows materials from a publication of the Ford Motor Company entitled "Continuing Process Control and Process Capability Improvement," published in 1987. Although most of the examples are hardware industry oriented, the course can be used successfully in process and service industries as well. Obviously, this requires capable instructors who can interpret and supply examples more readily understood by students in areas other than hardware. The use of relatively broad tolerances in the hardware industries may be difficult for students to understand if they are employed in some of the pharmaceutical, chemical, food, financial, or medical industries where close to zero tolerances are frequently the norm.

"Transformation" has been presented to staff members of a community college and was successfully used in team training to solve administrative problems (such as excessive student dropout rate early in the semester). Long after the course was presented, the college continued to use the team technique effectively to improve the quality of education and administration. It is an excellent method of introducing quality control techniques, and, with the availability of the 12 brief modules substituting for a textbook, students are encouraged to quickly and easily review steps already learned.

## Books, Magazines, and Journals

Whom can one believe? There is an abundance of books related to various aspects of *quality*. The American Society for Quality *quality press* publishes books written by over 300 authors. Each book has been peer reviewed, with frank opinions expressed in the ASQ's monthly magazine. However, only the favorable comments appear on the books' dust jackets, making it difficult to properly evaluate them.

We are all familiar with the deluge of books on politics and on dieting and the even greater number of cookbooks. An author with a novel idea proceeds to create a book to share the thought with the world. Some say you should eat no fat; others say fat is beneficial, but only if it is of vegetable origin, or of fish origin, or from the tropics. Some imply that a diet heavy in bran will overcome the ill effects of fat. The chances are that somewhere out there in the forest of books on fat, some author has probably set himself up as an expert on the all-fat diet. It is highly unlikely that each of these authors has found the secret to long life. They cannot *all* be correct.

The above is not intended to suggest that there is misinformation in the quality control books on the market. But, with the vast number of books on the subject of total quality, for example, it is entirely possible that there are some that have absolutely no application to a company in a particular industry or a company of a certain size. The 1994 catalog of ASQ Quality Press lists 17 titles that include "T" and "Q," or "total quality." Apparently, "total quality" must mean different things to different authors or there would be no need for such a proliferation of material on the same subject.

It has been said that, if nothing else, reading a book leaves an *impression* on the mind of the reader; it may be a precise impression of how to solve problems, or it may be a nebulous impression that vaguely suggests a new principle, a new thought, or a new field to explore. In between these extremes, there is a large assortment of mental responses. Unfortunately, it is a rare book that notifies the reader that he must have mastered certain prerequisites before starting to turn the pages. Some quality control books have innocuous titles but contain mind-bending formulas and contain as many Greek letters as English words. Most quality control books appear to be constructed to inform the reader of either a new quality principle or a new method of approaching an existing one. Others are popular subjects treated light-heartedly, more for entertainment than knowledge.

## Professional Society Courses, Lectures, and Seminars

The American Society for Quality offers a huge number of courses (138 in 1993) covering many types and levels of quality topics for professional and technical development. These are offered several times a year in various hotels and resorts around the country and are presented by recognized experts in the field. Should a company wish to have a group of employees attend, some of these seminars can be presented in-house, at a considerable savings. The course subjects and content are constantly updated. Some of the general fields of instruction are

- quality
- leadership
- engineering/manufacturing
- service
- information and analysis
- customer satisfaction

A hybrid type of quality control education is provided in many areas by local sections of national professional quality control organizations. For the most part, these are brief courses conducted by members of the organization to provide instruction in subjects not available at nearby universities or community colleges and are generally devoted to somewhat technical subject matter. An example of one ASQ section's offerings in a single year follows:

- Benchmarking
- Certified Reliability Engineer (CRE) Exam
- Certified Mechanical Inspector Exam Review
- Consumer Software Quality and the Law
- Certified Quality Auditor Exam
- CQE Exam Review
- CQT Certification Exam Preparation and Review
- Design of Experiments
- Geometric Dimensioning and Tolerancing
- Good Laboratory Practices
- Introduction to ISO 9000
- Introduction to Quality Assurance
- Introduction to Statistical Quality Control
- ISO 9000 Internal Auditor Training
- Job Search Techniques for the Quality Professional
- Medical Device
- MIL Specifications 9858
- New Product Introduction
- Process Validation Techniques for Medical Devices
- Quality Assurance Auditing
- Software Testing in the Real World
- Supplier QA
- Supplier Quality and Fraud
- Team Building
- Total Quality Management

This list of courses was offered by 18 instructors under the supervision of a 900-member section of the ASQ. Surprisingly, this program was successfully carried out in an area in which a community college was offering A.A. degrees in quality control. Apparently, there was a need for specialized education that was not available at the community college or elsewhere.

## In-House Quality Control Training

It is difficult to generalize on the subject of in-house training, but it is safe to say that very small companies provide their own basic quality control training in-house. The quality control manager takes the newly hired technician aside and hands him the company's quality control manual—assuming one exists. Detailed training is performed by the

senior quality control technician—assuming one exists. Further training is likely to be informal, consisting of questions and answers as the
need arises. It is expected that the larger the company, the more formalized is the in-house quality control training program.

In-house quality control training has a mix of good and bad attributes. In a small company, it provides an opportunity to train employees without removing them from their job site, thus keeping them
available should emergencies arise. By scheduling an hour or two
every few days or weeks, the employee's absence won't require a replacement during the training period. By avoiding an out-of-town trip,
morale might be improved because after-hours life is not interrupted.
Studying with fellow employees in a familiar atmosphere could speed
up the learning process. As apposed to off-site lectures to a mixed
audience, the training can be tailored to the specific processes of the
company, and the questions raised would have direct applications to
the students.

These are all potential advantages, but each has a downside. Students called out of class to handle a problem at their job site lose the
momentum as well as the specific information discussed during their
absence. A schedule arranged for the convenience of production might
be so intermittent as to require excessive review time at each session.
Some employees might resent losing the opportunity for an out-
of-town trip. Being in the same classroom with fellow employees
might stir up resentments caused by relative abilities or jealousy. Finally, the advantage of cross-fertilization of ideas from people in other
industries would be unavailable.

Assuming an in-house quality control training program is decreed, and
the subject matter, classroom location, and scheduling are established, the
selection of instructor(s) becomes critical to the success of the program.
Many large companies have full-time instructors on the payroll, complete
with classrooms and teaching facilities. On the other hand, their qualifica-
tions to teach quality control might be limited. In this case, the training
department has the opportunity of selecting instructors from within the
quality control department or looking to the outside for university profes-
sors, professional consultants, or seminar instructors.

## In-House Training by an Inside Trainer

In-house quality training in a large company presents other opportunities. If a company's formal quality control training program has been

in use for years and is required attendance as part of indoctrination of new employees, there is rarely any emotion expressed. When a new quality control training program is set up within the company, using a company employee as trainer, a wide assortment of responses arises, usually made to a fellow employee, some favorable, many of them brutally frank:

1. "I'll do it because I have to." (This is part of the job, and, if I don't behave, I'm liable to get transferred, or fired. At least, while I'm here, I don't have to listen to all those ringing telephones.)
2. "When did *he* (the trainer) get so smart in quality control? Those guys in the personnel training department might be great instructors in accounting, but all they know about quality control is what's in the videos they're showing and the book they're reading from. I can do that much without him."
3. "I've been doing this job for 10 years; what more does *she* (the trainer) know about it?"
4. "This is another management idea. It, too, shall pass."
5. "Nice theory, but you know it (empowerment) won't work here. I tried to change the system a couple of years ago and got into big trouble over it."
6. "We've requested expert help—look who we get!"
7. "Why can't I just put in my eight hours and go home to my six-pack and TV?"
8. "You just don't understand the problems of the filler–sealer."
9. "The Union *said* they would go along with this, but you just don't know Charley, the shop steward."
10. "Will you go to bat for us if we need help?"

## In-House Training by an Outside Consultant

1. "Good stuff, but my boss doesn't believe in it."
2. "If I make it work, the boss will look bad."
3. "I'm not allowed to discuss that (sacred cow)."
4. "The accounting department will never give us that information." (turf protection)
5. "My boss has an MBA; I never finished high school. Why would he listen to me?"

6. "After you started to teach TQM, they dismantled the TQM department; what does that mean?" (It probably means that management is not serious about TQM, and that this course will be meaningless.)
7. "The whole company has been told to make it work; it must be good." (This is the type of message that every manager wants to hear. It suggests that the training program has been successfully advertised to the employees.)
8. "I've always said our scrap was too high; here's the chance to fix it." (More good news! The instructor has apparently hit the motivation nerve.)
9. "How do we get our preprocessing department to give us good material to complete?" (A good indication that this employee has learned one of the concepts of quality control.)
10. "We've never used p-charts. How do we sell the idea?" (Also an encouraging sign that this course is making progress.)
11. "Great! Now I'll know what to discuss with the quality engineers." (another sign of success)

## OBSTACLE NO. 5: PEOPLE PROBLEMS

Not surprisingly, the most serious obstacle to quality control training is people. (There will be more on this subject in Chapter 3.) Some of the preceding discussions have assumed that all people were exactly alike, that they were all equally eager to learn about quality control, that they were all clever, open minded, serious, healthy, dexterous, quick to learn, and all had equally high IQs. This is obviously nonsense, and yet it is universally overlooked when selecting candidates for training and frequently when devising lesson plans. In one series of four classes of line supervisors, roughly 20% were above normal in their ability to handle algebraic expressions and were close to sleeping through a basic review mathematics course. Seventy-five percent managed to absorb the fundamentals, but 5% could not be convinced that multiplying a positive integer by a negative one would result in a negative answer. With classes of this composition, how does one design a lesson plan for calculating standard deviations that will avoid reactions varying from utter boredom to downright fear? If a company sends some line employees off to learn statistical quality control, will they return with new knowledge, or will they return baffled and distressed?

It has been reported that one of Germany's finest automobile manufacturers trains high school dropouts and other young people in machine shop, metrology, and assembly. The training carries these workers through all of the steps of the trade with hands-on practical experience. One of the company's major concerns is that, after the training is completed, the trainee will take all of these skills elsewhere, rather than remain on their assembly line. The same thought must pass through the minds of American managers when there is an opportunity to send an employee to a quality control course.

This has become an increasingly difficult problem as employee mobility has been on the rise. Although the following numbers are subject to considerable interpretation, they suggest how common job mobility has become. Around 1988, an employment agency manager stated that, on average, salaried employees changed jobs every six years. In 1998, a job-search company official stated in an after-dinner speech that salaried employees changed jobs every two to four years. In evaluating these two facts, it may be important to note that one agency was on the East Coast, and the other was in the western aerospace area. It is perhaps more important to realize that not everyone who decides to remain or move on changes jobs through agencies. Yet, it is a fact that even without "downsizing" and recessions, employees do lose their positions and are forced to look for new ones at an increasing rate. Others become dissatisfied or bored with their jobs and look elsewhere. The days of lifetime careers with a single employer are probably ending. Fortunately, this is a two-way street. For every trained employee lost by ABC Corp., there is a distinct possibility that he will be replaced by an employee trained by XYZ Corp.

Increased mobility raises another concern: trade secrets. Small companies are more likely to retain trade secrets than larger ones, since the concept is if they're small, they couldn't be doing anything spectacular. Therefore, if an employee moves from a small company to a larger one, it is doubtful that the larger company would be interested in asking him about the smaller one's secrets. By the same token, the smaller company might believe that they would be unable to afford to use the secret processes of a larger company because of the supposed need for unaffordable larger equipment.

Secrets leak out of companies so easily that it is surprising that any are left. The company that applies for a patent is protected until the patent is granted. Surprising? Not at all. As long as the patent is still pending, competitors are unwilling to chance an infringement lawsuit. Consequently, the words "patent pending" on a competitor's product mean "hands off" until the details of the secret process are published in a granted patent. At that time it is often possible to find a way around the patented process through slight modifications that are not covered in the patent.

With the advent of supplier certification and the accompanying supplier quality partnering over the past several years, it has become increasingly difficult for both supplier and purchaser to retain company secrets. If a large company offers to help its supplier with a quality program, the large company's competitors may benefit from the improved supplier quality as well. Additionally, if the supplier is to understand the needs of its large company customer, in a partnering arrangement, the supplier must learn of the customer's operations in some detail. There go the secrets!

There are many puzzles associated with partnering. Suppose the customer solves a quality problem for a partner supplier. Is the customer required to give the same information to another supplier of similar products? If a small supplier's partnering customer has a process problem involving purchased materials, is the supplier obligated to help by offering information acquired from another customer—possibly even a competitor? There seem to be more questions than answers.

Over the years, in an effort to keep trade secrets safely within a company's borders, employment contracts used to be commonly drawn up for quality control engineers. In the rare cases where a lawsuit resulted from the engineer's moving to a competitor, along with the secrets, the lawsuits generally ended in favor of the engineer. The reasoning was that the engineer had every right to earn a living based upon his training and experience, provided the information he took along with him was in his head, not on his ex-employer's memoranda.

In trying to protect company process secrets, one company removed the scale faces from pressure and temperature gages, replacing the numbers with letters: A, B, C, etc. to replace 100, 125, 150, etc. This provided the secrecy until one night when somebody from the maintenance department met a competitor at a bar by pure coincidence, and they started to compare notes!

One of the most closely guarded secrets in the spray drying industry is the specification for the nozzle. Out of the thousands of designs available, there is usually only one that has the ability to produce spray patterns, flow, velocity, and size that meet the quality requirements of the processor. Imagine the astonishment of the construction engineer when the owner of a nearly completed drying plant arrived one morning and handed the engineer a duplicate of their major competitor's nozzle. How did he find it? Perhaps from an employee. Perhaps from the nozzle manufacturer. Perhaps from another competitor. There just aren't any real secrets.

Why all this discussion about trade secrets? The answer is simple: if the company sends its engineers and quality employees off to a seminar or classroom full of competitors' technical personnel, will they inadvertently share company secrets with each other? This is at the heart of the trade secret obstacle related to quality training.

Why are there turf problems? Employees who do good work are proud of their accomplishments and wish to protect their domain (turf). Employees who are on shaky ground with their accomplishments are equally anxious to protect their turf. This usually manifests itself in fearful, even angry or misleading remarks such as the following:

1. Quentin the quality manager to Pete and Mary, line operators:

   I'm thinking of sending you off to a week's classes in statistical quality control.

   Pete: Can't you send Phil? He's good at that stuff.

   (What Pete means: Why? I never did like mathematics.)

   Mary: For eight years it worked OK my way. Why change?

   (What Mary means: I don't want anybody stepping into my job and showing me up while I'm gone.)

2. General Manager to Quentin: Why don't you send Charlie and Marie to a statistical quality control seminar?

   Quentin: They're pretty busy right now. Let's talk about this next month. (What Quentin means: No employee of mine is going to tell me how to operate my line better.)

On the positive side, those employees who have a chance to further their quality education at company expense frequently return to their job a little prouder, a lot smarter, and with a better understanding of their quality contributions.

## SOME SOLUTIONS

To start with, let's agree that there are some training problems that have no solutions. Some of the sticky situations discussed in this chapter may not be solved for years. Others have been resolved fairly quickly. Following are some suggestions for alleviating, if not solving, those discussed.

### Obstacle 1: "Quality" Definitions

There are as many definitions as there are definers. The best advice is to select one that matches your company's products and goals. However, be sure to separate "quality" from "quality control." If you're in upper management, tell your divisions what it means; if you're in middle management, suggest your definition to upper management and modify it as necessary, but don't work without a definition; if you're on the line, ask your supervisor what it means, and support it.

### Obstacle 2: Differing Company Quality Goals

For quality control to function, everybody in the organization has to know what it means. Upper management should announce an unambiguous policy; if none is available, quality management should send a tactful inquiring memo to upper management with a suggested quality policy statement. If wrong, it must be corrected and incorporated into the quality manual as a cover sheet. Line personnel should understand the balance between production quantity and quality required by the supervisor.

### Obstacle 3: Buzzwords

Brace yourself for new buzzwords every year or so! Check them out so that you can determine if there is really something new that will be of value. If necessary, create your own buzzwords. Select a few and try them out for acceptability upward, crosswise, and downward through your company. Stay with it for a year or so or until its effectiveness wears off. By that time there should be new ones in the quality control literature. Ensure that those *to* whom you report, those *with* whom you report, and those *who report to* you all have the same understanding of the term.

Simple charts and graphs can be very useful for awhile. Use them sparingly, and make sure that everyone in the chain of command from the top to the bottom sees and understands them. After a few cycles, stop publishing them and see if anybody notices. If so, try an "improved model" for a time; if not, stop for awhile.

## *Obstacle 4: Confused Growth and Availability of Training*

How do you check out the speakers, instructors, and seminar leaders?

Generally the brochures describing the course of instruction will include brief biographies of the speakers. Look for real-world experience, preferably in companies the size of yours. Read the testimonials with a grain of salt, but take note of who submitted them. If you're really serious about understanding what the program is all about, and the value it might have for your company, contact a couple of the testimonial writers, and question them.

How do you select the course? With the hundreds available, the course should match the level of the attendee's position in the company.

- Top management should concentrate on courses in total quality control, and quality planning, with some exposure to quality assurance, product development, and team concepts.
- Engineers and middle management should add quality control techniques, statistical methods, and product development.
- Line employees should receive elementary instruction in quality control techniques, statistical methods, and team concepts.

How can you check out the course content? Contact the course presenter and ask for details. Then ask for phone numbers of previous attendees. Call them and find out (1) if they learned anything worthwhile, (2) if they were able to apply it in their company, and (3) if it did any good.

Better still, attend monthly meetings of professional quality control organizations, and talk to the people who are using the techniques of interest to you. They can provide information about training sources as well as the advantages and disadvantages of different quality methods.

## *Obstacle 5: People Problems*

How do you select the right candidate to attend a seminar? Know the employees' capabilities in the planned field of instruction. Then, and only then, can the employees and the subject matter be selected. Problems may result from poor planning. For example, setting up two sessions of one hour each to train the accounts receivable clerks in construction and application of Pareto curves is likely to result in chaos. Assumptions here must have been that anyone who deals with numbers (such as accounts payable personnel) must understand how to conduct sampling plans to collect data and how to prepare, graph, and interpret cumulative averages. These assumptions are unlikely to be correct.

Trade secret protection is an annoying fact of life. The secrets will probably escape eventually, but at least try to delay this. On the other hand, don't hide them to an extent that the employees who really would benefit from knowing are kept in the dark. ("You mean that threaded hole is a pressure vent? Nobody told me. I've been screwing in a machine plug all week.")

In the following chapter, we will explore the obstacles surrounding the implementation of a quality control system.

# CHAPTER 3

# Implementing Quality

## THE QUARTERLY FINANCIAL REPORT

PERHAPS the greatest obstacle to quality and process improvement is the shortsighted planning span of some upper management. It may not be entirely their fault, since their performance is judged by the financial report generated every quarter. Any demonstrated improvement proposal in any phase of a company's operation will invariably be accepted and implemented immediately by management—but only if it has no adverse effect on the quarterly report's bottom line showing net profit. This is possible under any of four conditions:

1. The improvements will save costs of labor, materials, etc.
2. Installation of the improvements has no associated costs.
3. The cost of the improvements will be immediately offset by additional profits within three months.
4. Failure to install the improvements will cause a loss of profit over the next three months.

This is obviously somewhat of an overstatement. It generally takes more than one undesirable quarterly financial report to cause the demise of an executive. There are also extenuating circumstances that may negate the overall importance of the report: mergers, overall market deterioration, takeovers, major national economic developments, changes in governmental regulations, and physical disasters such as hurricanes, floods, fire, or earthquakes. On the other hand, the importance of the financial report cannot be underestimated. The emphasis on short-range evaluations has resulted in short-range planning.

Fortunately for those involved in quality control, there are usually enough of these improvements ready to be uncovered and installed to guarantee job security. More often, however, there are opportunities to make appreciable contributions to quality and profitability that do not fit into the required quarterly time frame. A few examples include long-range projects that require new machinery, additional processing facilities, major changes in raw materials, additional floor space, modification of equipment layout, or additional labor. In the face of the threat of the quarterly profit need, how can long-range quality improvements be presented in such a way that they may be favorably considered?

Following is an example in which the quality control manager at a small company has a profit-generating idea that will require an initial outlay of $50,000 for an analytical laboratory device. This proposed analyzer will permit vastly improved control over the quantity of an expensive ingredient required in a process, saving a modest $10,000 the first year, but increasing the potential savings geometrically as the company's volume expands.

Attempting to purchase an analyzer that will cost the company a net $40,000 the first year would have been a lost cause. Since a request for outright purchase doesn't have a chance of success, the quality control manager investigates equipment-leasing organizations and finds that this analyzer may be lease-purchased on a five-year contract for $15,000 per year. Now, the first year's loss is $5000 ($15,000 lease minus $10,000 savings). Knowing that the future holds great promise of savings, he now searches for a means of demonstrating this program in a favorable light, and decides that a picture would have a better chance of success than a memo. With the help of the accounting department, numbers are generated and graphed as shown in Figures 3.1 and 3.2.

Table 3.1 seems to show that an outright purchase would result in a

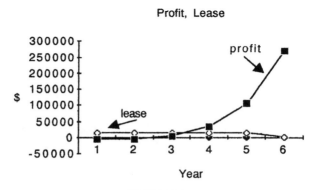

**FIGURE 3.1.**

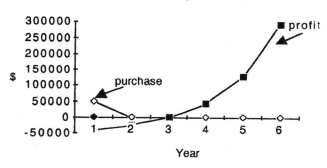

**FIGURE 3.2.**

net loss of $40,000 in the first year and that it would take another two years merely to break even. Table 3.1 does not show that, although the balance sheet would look unpleasant the first year, the equipment would be paid off at that time and that a savings of $15,000 would appear the second year, followed by a larger savings of $25,000 the next year.

The lease option looks a bit better. The net loss the first year is only $5,000, and the proposal starts to show a profit after the second year. The obvious drawback to leasing is that the equipment would cost $75,000 instead of an outright purchase price of $50,000. This may be offset by a more careful analysis of the time value of invested money.

Table 3.1 does show great profit potential from either choice; however, graphing these data does not show a particularly startling difference between the alternatives.

*Table 3.1.  Estimate of Savings over Six Years*
*Using New Testing Equipment ($).*

| Year | New Processing Savings — Estimated Savings | New Processing Savings — Cumulative Savings | Outright Purchase — Purchase | Outright Purchase — Cumulative Savings | Five-Year Lease — Lease | Five-Year Lease — Cumulative Savings |
|------|-------------------|-------------------|----------|-------------------|--------|-------------------|
| 1 | 10,000 | 10,000 | 50,000 | −40,000 | 15,000 | −5,000 |
| 2 | 15,000 | 25,000 | 0 | −25,000 | 15,000 | −5,000 |
| 3 | 25,000 | 50,000 | 0 | 0 | 15,000 | 5,000 |
| 4 | 45,000 | 95,000 | 0 | 45,000 | 15,000 | 35,000 |
| 5 | 85,000 | 180,000 | 0 | 130,000 | 15,000 | 105,000 |
| 6 | 165,000 | 345,000 | 0 | 295,000 | 0 | 270,000 |

From the above discussion, it would appear that the best argument is the outright purchase of the analytical equipment, showing the entire cost the first year, and then enjoying the profits thereafter. The graphs assist in demonstrating the projected growth of profits and should soften the blow to the first year's balance sheets.

The lesson from this hypothetical experience is that quality control managers need to look at the plight of upper management when presenting a costly project for approval. The quarterly financial report obstacle cannot be overlooked during the presentation, and it is wise to prepare in advance for an initial negative response to a proposal.

## PEOPLE: THE SECOND MOST SERIOUS OBSTACLE TO QUALITY IMPROVEMENT

There is always room for argument regarding the worst obstacle to quality control. The quarterly financial report has been labeled the number one obstacle since it is a cold, inanimate object and is extremely difficult either to avoid or change. On the other hand, although people may appear to be inflexible in their beliefs or behavior, it is frequently possible to nudge them in one direction or another.

Studs Terkel, author of *Working*, wrote of his interviews with over 500 workers in a wide variety of jobs. They spoke freely of their joys, sorrows, problems, and frustrations, but the reader would find it difficult to find suggestions for making things better. These personal interactions are not limited to manufacturing or service. Dr. Gerald Edelman commented that there was "very often a strict hierarchical structure in science. The head of a lab could be a very scary paternal figure, loved and hated. I learned that there was an awful lot of backbiting and fierce competition over small matters."

Self-centered or abrasive bosses at any level of the organization can generate severe morale problems among the workers, invariably resulting in loss of productivity, increase in mistakes and defect production, and ultimately an adverse effect on profits. If a clerk approaches her supervisor with a suggestion for improving the quality of customer service and is told, "When I want your advice, I'll ask for it," there is little likelihood that the clerk will ever again try to improve performance. Consider an assembly line operator who suggests to the foreman that it might be more efficient to tack on the wiring harness before installing the computer chips instead of the method now in use. If the response from the supervisor is, "Just do as you're told, and quit bothering me with your stupid ideas," would anyone else in that department ever be expected to come up with another process improvement suggestion?

Perhaps abrasive company policy falls into the same category. Throughout the mid-1990s, companies went through a period of restructuring, downsizing, reengineering, or just plain laying off, in which as many as 40,000 employees from one corporation were eliminated from the payroll in a single year. The chances are slim that either employees or managers will place much emphasis on quality under these conditions. Survival would be the key concept—survival of job, survival of company, survival of customers. It is not difficult to visualize the creation of the quality scapegoat when major cuts are made. When things go wrong, fire the quality manager (not the plant manager, not the purchasing manager); we don't have to have him.

The existence of manager tyranny is widespread. Several books have been published dealing with the subject, and there have even been national contests to select the year's worst boss. One author has labeled the offender as the "Toxic Executive" and has written a book on the subject, along with suggestions for possible cures. Some of the worst examples of bosses who thrive on conflict:

- gives explicit instructions for a task and later ridicules the employee for following them
- in the midst of a crisis, issues several conflicting demands, thus creating additional crises, blaming those involved
- criticizes an employee in the presence of fellow workers
- searches a worker's desk, e-mail file, and locker, looking for evidence of conducting personal business on company time, and then confronting him with his suspicions
- always severely admonishes a worker for his mistakes but never, never compliments him on a job well done
- insists (often with contempt) on needless repeated corrections, modifications, adjustments, improvements, perfections, and amendments to a completed task

Most of us harbor a bit of a Walter Mitty dream world. (He's the fictional character who would drift off into a reverie wherein he was the invincible hero of a situation.) Were there not a few times when we visualized unlikely ways of getting rid of an obnoxious or competitive fellow worker or a stubborn, mean boss?

This is a sick, inhumane thought. Why bring it up? Because this book is about obstacles to quality control as well as solutions and, all too often, at any level of the organization, a serious obstacle is the boss, a fellow employee, or an underling. Fantasy is one thing, but cruel and cunning ploys have actually been tried against a perceived "enemy." Sometimes they work; sometimes they backfire. They invariably damage the character of both the accused and the accuser.

The Walter Mitty in us rubs his hands in glee as he proceeds to plant seeds of mistrust in every corner of the organization, both above and below the boss's level. If Charley the boss is of advanced age, "Don't you think that Charley has outlived his usefulness?" If he is fairly young: "Seems like Charley has run out of fresh ideas lately." How about "Charley is always making mistakes. Keep an eye on him." If he's always cheerful and smiling: "Too bad Charley is always fooling around. Why can't he take his job seriously?" If nothing is that obvious, try "Charley is always negative."

A more genteel approach to get rid of Charley would be to send a glowing report anonymously to an executive search firm describing Charley's best attributes, along with a number of good ones he ought to have. Goodbye, Charley! He's off to greener pastures.

Now that we have fantasized about "getting even," let's look for more acceptable ways of getting *around* the boss, or getting *along* with the fellow worker, not getting *rid* of them. We cannot expect to find a single universal cure. However, here are a few suggestions.

- Befriend someone who is close to the boss, and route ideas through him.
- Send him birthday and anniversary cards (to his home); show him you care
- Find some reason to compliment your fellow worker: performance, an idea.
- Do not attempt to tell him how clever and educated you are; let him find that out for himself. At the same time, keep in mind that, if you place a low value on yourself, be assured that the world will not raise the price! (This is a sensitive balance.)
- Ask for his suggestions and aid. Don't broadside him with a proposal.
- Out of the blue, offer some sympathy about something, then walk away. This might set the stage for a friendly reception when presenting a proposal.

- Divide your proposal into three parts, and present one third of them over three days. Move cautiously. Bear in mind that you are far more familiar with your corner of the business than he is. Don't discuss page 4 while he is still trying to absorb the third paragraph on the first page.
- Give credit for your idea to a friend (real or imagined) who works for another company in another industry, concluding with, "Do you think that idea might work here?"
- When all else fails, ask for a heart-to-heart talk. Your fellow worker or your supervisor will likely present you with some troubling information about your proposals or your methods of presenting them. There is no point in defending them, but much will be gained by acknowledging the shortcomings and agreeing to make whatever changes are indicated.

Finally, following are some actual examples of real individuals and their relationship to quality control—as gleaned from interviews with dozens of professionals in the field. This list of people problems (some with solutions) is not intended to be complete, but perhaps a useful pattern might be found in this collection.

### Characteristics of People Obstacles

H is a likable, though totally self-satisfied, blowhard. He likes to invent situations in which he appears to be well educated (which he is not). He is in charge of three quality control operations that are in a highly technical field, of which he is ignorant. How could a person retain his position with so little competence? Probably because he is very personable, a superb plotter, an accomplished politician, a convincing actor with the gift of gab, and he has surrounded himself with capable people. Any complex idea from his quality control people is immediately quashed; any good idea that he does understand he snatches to his bosom as his own concept. How can his three quality control managers do their job under these conditions? For the most part, they have quietly operated within the guidelines already existing, taking care not to "rock the boat." Whenever a novel idea comes up, they are careful to present it to H in such a way as to allow him to "think of it himself." If there are technical aspects to the idea, they explain these very carefully as they arise. Has this approach worked? If "worked" means "has everyone kept their jobs?" then the answer is yes. Has quality control effectively improved quality and productivity? The answer is probably not.

M is a well educated person. She knows her quality control job control, and furnishes upper management with a continuous flow of memos and reports that show steady progress. She is somewhat weak in providing leadership and rarely solicits suggestions or directs the technicians who report to her. Once in a while a technician will approach her with a quality control improvement suggestion, only to find it languishing in her "pending" file weeks later. Eventually, the technicians found that, if two of them approached her simultaneously with a suggestion, she seemed to be reassured that the idea had been thoroughly thought out, and reacted quickly, almost as if she thought she was "outnumbered."

PS, the plant manager, is the rare bird. He is bright, thoroughly experienced in his field, and eager to assist younger employees. He is not an obstacle to the advancement of quality control in the company; however, the quality control manager might be his own obstacle. With a man such as PS, the introduction of an improvement in product or process had best be thoroughly thought out by Quality Control before presenting it, or else there is a good possibility that PS will ask some unanswerable questions, thus killing the proposal.

G is dishonest. Unfortunately G owns the company, and you report directly to him. G is the kind of person who will arrange to buy floor sweepings and other companies' rejects as the raw material for his food process, stating that the customer will never question the quality if the product *looks* good. This man is not just an obstacle; he is a disaster waiting to happen. There is only one satisfactory course of action in such a case: resign. He cannot be "cured," and, if you remain, you will become contaminated.

L is a thoroughly credentialed engineer, and a fine gentleman as well. More important, he is a major stockholder and director of the company. He is also very wealthy, and the concept of time is unimportant to him. Two employees report directly to L: E, the plant engineer, and Q, the quality control manager. E has been with the company since he was a teenager, and now, in his early 60s, his only goal is to remain innocuous until retirement. As a result, he gets along very well with L, although he accomplishes very little. Young Q, on the other hand, is engaged in a major process change that requires some heavy-duty engineering. Because E is of no help whatsoever, Q bypasses him, dealing directly with L. Each proposal is scrutinized by L, and he insists on constant study and restudy of foot-pounds, air flow, energy, and the effect of rotation of the earth at each step. The project is becoming bogged down, and E doesn't care. Q is becoming frustrated. Is there any reasonable way around obstacle L? As it turned out, every few weeks, the company president would ask Q how the project was coming along. By very careful wording, Q was able to convey to the president that progress was slow, but that there could be complete assurance that it was in the right direction, since L was so very, very, *very* thorough. Eventually the message came across, and the president arranged for L to take a three-month vacation in Europe, during which time Q managed to successfully complete the project.

P is friendly but shallow. She has been with the company for several years, and her relation to the CEO appears to be somewhat more than businesslike. Over the past few years, she has been promoted from department to department and is now in charge of purchasing. Fortunately, there are several very capable old-timers in the purchasing department, and things appear to be running smoothly. One day, after making a series of studies relating the costs of production to the quality of raw materials, the quality control manager drops in to visit P and discuss the possibility of being a little more selective of the shipping case supplier. Quality Control points out the number of rejections and packaging line stoppages that are the direct result of defective cases, and attempts to demonstrate how the costs of improved case quality might offset the cost of scrap, rework, and downtime. In an explosive temper, P shouts at Quality Control, demanding that he mind his own business, and get out of her office, and "go run some tests or something." Days later, while discussing this disaster with another department head, Quality Control discovers that P has been the recipient of numerous dinner and theater evenings and other gifts from the present shipping case supplier. Obviously, Quality Control should have approached P in a less direct and more exploratory fashion and perhaps might have tried to learn more about how P operated the department before offering suggestions for improvement. In the future, Quality Control will have to spend some time and effort in patching this torn relationship.

S is a handsome, personable, intelligent glad-hander whose philosophy of getting ahead in business is to "tell them what they want to hear." He is brought into the company with a somewhat nebulous position, and after a few months replaces the general manager. After S has rearranged the organization, Quality Control finds himself reporting to

S, rather than to the company president. Unfortunately, there are no successful quality management philosophies that are based on reporting only the good news. This is a very painful relationship, with few options leading to a successful quality improvement program. Quality Control attempts to meet S head-on by presenting some cold facts about the control of product quality, with no success. He then tries to go around his boss and present these facts (and suggested improvements) to the departments involved. This meets with stern disapproval from S, along with warnings about longevity of employment. Is there a way around this obstacle that will permit Quality Control to operate effectively? Quality Control gradually evolved three partial solutions:

1. Always tell S "what he wants to hear" before coming up with less pleasant quality-related news.

2. When the news is bad, release only tiny fragments of the unpleasantness, along with liberal praise for whatever good news can be found, and with suggestions for maintaining the good performance. Quickly follow up with more of the same until the entire story is transmitted to the department affected.

3. Bring any proposed quality-related memo or report that has even a trace of unpleasantness directly to S and request his suggestions for wording.

Vice president Z is a close friend of the president, and very effective in his job. He has a wide network of friends in the business and constantly brings interesting new ideas to nearly every department. His one failing is that he frequently enjoys "liquid lunches," returning to work somewhat inebriated. On many occasions, he provides slurred unintelligible suggestions to one department or another, which more or less are ignored. Not so with Quality Control. When Z presents Quality Control with a blurry proposal for a test run to improve quality, he calls for results the next morning. Worse still, he then asks why Quality Control ran such an idiotic test yesterday afternoon, forgetting that it was at his request. Drunken behavior in low-level jobs is easily taken care of through the personnel department or the shop steward. But a vice president? Quality Control thinks about the possibility of confronting Z directly, or even the president, but quickly realizes that the obstacle to be removed would probably be himself. Eventually, Quality Control tries a successful plan, wherein he dutifully copies down the afternoon's drunken suggestion and does nothing with it that day. The following morning, he seeks out Z and discusses the notes, asking for guidance on one or two points before conducting the test. Z will not admit that he doesn't recall these soggy suggestions and finds a convenient reason to postpone or cancel the test. After a few months, Z abandons the practice of approaching Quality Control with after-lunch test runs.

RD [the research and development (R&D) manager] had always claimed ownership of the quality control function and was angered by the abrupt appearance of a new quality control manager. He made no attempt to hide his displeasure, dropping unpleasant remarks about Quality Control whenever and wherever he could. The jealousy obstacle frequently clears itself out of the way over time because it is so obvious to most other employees. On the other hand, a new quality control manager might not be blessed with much time to demonstrate his abilities, and steps should be taken to eliminate this obstacle as soon as possible. Direct confrontation rarely works when emotions run high nor is it likely to expect assistance from, say, the personnel department. This is one of the few times one has to complain directly to father. Under these circumstances, "father" refers to the quality control manager's supervisor, or R&D's superior, or perhaps both.

C was the vice president in charge of a division. He was, in a word, weird. He talked funny, he acted strangely, he moved in jerks, and he dressed peculiarly. Quality Control arrived at the division office one day after an interview at the company headquarters in another city, with instructions to learn all there was to know about the company over the next 52 weeks, in preparation for a major position somewhere in upper quality management. Either nobody bothered to tell this to C, or C totally forgot about it during the year. The result was that, after one year, C called Quality Control into his office and fired him. Now that is an obstacle! As it turned out in this case, Quality Control was not comfortable with C's management or personality and graciously accepted termination pay and left. One would expect that, if Quality Control had remained interested in the initial offer, he would have taken steps to discuss his progress with C many times during the year of trial employment, thus preventing this insurmountable obstacle from appearing.

O was a young lady who had worked for the engineering department for several years as one of the secretaries. When a new quality control manager made his appearance at the office, Personnel decided to attach O to Quality Control as his secretary. Within a day or two, Quality Control realized that O was trying to act like "one of the boys," and constantly employed an extremely tactless choice of coarse words. Quality Control strongly believed that anyone involved in quality control should display an aura of personal "quality." Realizing that O's crude behavior was an obstacle that would adversely affect the quality image, he requested that personnel return her to engineering and find a replacement for the quality control department.

Company president J greeted the newly hired quality control director on his first day of work and explained that he had neither time nor interest in the operations of the quality function. He assured Quality Control that he would fully support any of his decisions but that he would prefer to be left out of the loop, since he was very heavily involved with company expansion and finances. Quality Control failed to recognize the seriousness of the trap that J had set with this philosophy and proceeded to take it as a directive to act alone, not bothering J with his progress or problems. Thus, Quality Control prepared his own obstacle, which subsequently led to his downfall. Periodically, a couple of jealous old-timers in R&D and Production would bend the ear of J with unpleasant observations regarding the new quality control director. This one-sided campaign led to Quality Control's finding himself in hot water. Should he have seen this coming? If Quality Control had set up an adequate intelligence network, the answer is "yes." In any event, Quality Control most certainly should have attempted to keep in touch with J until he learned of the jealousies and worked out a strategy to combat them early on. If it turned out that J really meant that he literally did not wish to be bothered with quality control matters, he would most certainly have reminded Quality Control again.

B is the third- or fourth-generation family president of a fairly successful process industry company. Although a business major, he seems to lack experience, drive, and direction. Worse still, he is impulsive and unpredictable, generating business decisions seemingly at random, often conflicting, and without solid data. In short, he is a king-sized obstacle. Following is an example of a quality control dilemma. A series of test runs showed that many of the serious quality problems were caused by line changeovers and short runs. Quality Control generated a proposed longer-range production schedule and presented it to President B for approval. It was enthusiastically endorsed and was put in place the following week. Results were even better than predicted. One month later, without explanation, President B instructed production to "stop this schedule nonsense" and go back to the old system. The quality defects reappeared. Quality Control thought it wisest to say nothing for the time being. Perhaps, after a few more months, he might bring the data to B in the hope that he would be in a receptive mood and would reinstate the newer system. On the other hand, B was equally likely to fire Quality Control on the spot. Perhaps Quality Control should have been presenting the favorable daily data to B during the few successful weeks, but, considering the sometimes explosive and generally unpredictable nature of his boss, Quality Control could not be faulted for the inaction.

Q was promoted to head one of the quality control groups. On his first day, the boss told him, "Be firm with your crew. If you're not firm, they'll walk all over you. If they walk all over you, I'll find someone else for your job who knows how to be firm—and you're outta here." A few days into the job, the boss offered Q some more advice. "The human resources department is giving a one-day course for new supervisors. They are going to tell you that your crew's job should be *productive, constructive,* and *fun.* Don't pay any attention to that *fun* stuff. We're here to make Finnegan Pins; if they want fun, they can do it on their own time." How could Q remain on good terms with his boss, the human resources department, and his crew and still do his job?

Without knowing the people involved, it would be difficult to suggest a solution. Q decided to find a convenient time to discuss progress in his new position, during which he told the boss that he had called his crew together and explained what he expected of them. Q did not mention that nobody in the crew thought to ask what "firm" meant. Nor did Q say that he neglected to use the word "fun" in his discussion. Fortunately for all, Q's thoughtful plan and tactful explanation succeeded in creating a good working relationship with the boss.

### How to Tell When Quality Control Efforts Are Endangered

A people obstacle (and a very painful one) is when the quality control employee suddenly experiences a jolt, as if a rug had been pulled out from under. This can happen to anyone, with or without justifiable cause. In the case of quality control, it generally appears during a period of job euphoria—everything is going along smoothly; three successful projects have been completed, and, when installed in the plant, they work perfectly. Unfortunately, euphoria sometimes causes myopia. We are so consumed with the success of the moment that we forget to look around and observe changes in the environment.

In the event that somebody has started a rumor that Quality Control is "outliving its usefulness" or that you're "running out of fresh ideas lately," chances are that the first indication will be some guarded sympathetic glances your way from employees you never really knew well.

A second indication might be the unexpected offer of assistance on one of your projects by an employee at your level of management. In this case, remember the homely advice: the only free cheese is in a mousetrap.

A third indication is in the form of an impossible request from upper management for an immediate report showing a detailed solution to a problem that will take three years to explore, much less solve.

Fourth are vague statements at increasingly frequent intervals about your abilities, from the boss and his assistants, like "you really should be more of a team player," "your reports are getting too long (short, complicated, infrequent, etc.)," or "the marketing manager asked me if you knew why we lost the Jones account last week."

Fifth is an informal request from the boss to be in his office at 10:00 A.M. to discuss your plans for next year—or some other meaningless long-range subject.

This is not just another obstacle. We may be looking at a disaster here. There are three possible remedies:

1. Put your spy system to work and find out who is behind the vendetta, and try to talk it out.
2. Approach your boss with quiet self-assured dignity, and ask how you might improve your effectiveness.
3. Look for another position.

### Incoherent Growling Leadership

Several years ago, the shop foreman was characterized as a cigar-chomping, grumpy slave driver, who spoke in short unintelligible gruff snorts. He was dressed in baggy black pants, a vest with gold watch chain, and rolled-up shirt sleeves. Over the years, there has been considerable humanizing of this incoherent type of supervisor. Unfortunately, some vestiges remain, although they have mutated to a form of communicating characterized as "mumbling."

The purpose of mumbling is not to confuse, but to present an aura of dignity, command, knowledge, and superiority without actually adding any useful information to a directive or to the solution of a problem.

Newspaper comic strips have poked fun at this type of shallow leadership. A book on the subject, *When in Doubt, Mumble,* by James H. Boren, is an amusing account of the techniques of mumbling and contains many principles of interest.

One of the common methods of obfuscation described in the book is a method of selecting meaningless descriptions of a process or technique by combining groups of three words (adjectives, adverbs, and nouns) each of which has a distinct meaning but when used together have no significance. A single multi-syllable word is selected at random from each of three columns. The three words may be forcefully spoken, but mispronounced, or they may be spoken quietly and rapidly: mumbled. The book provides full pages of lists that are grouped for use under different circumstances. A few illustrations of some of the types of words are shown below:

| #1 | #2 | #3 |
|---|---|---|
| Documented | Physical | Procedure |
| Computerized | Analytical | Allocation |
| Modernized | Variable | Implementation |
| Optimized | Hypothetical | Presentation |
| Challenged | Minimal | System |
| Applied | Orbital | Guideline |

Following are some examples of how the mumbling system is used:

- Example 1: (in answer to an employee's question about the need to keep certain handwritten forms) It is important to keep production records on form #37A-6 because it provides a *documented minimal allocation* of the data.
- Example 2: (at the end of an admonishment to a retort peeler operator who has just created several overcooked batches) If you can't keep the temperature within tolerance, we're going to have to prepare an *optimized variable procedure* for you to follow.
- Example 3: (introducing the subject at a department meeting) I've been informed that management is disappointed with our output last week. When you go out onto the production floor this morning, I want each one of you to constantly remind yourself of the importance of using an *applied analytical system* to your work.

How do you, the quality control manager, respond to a production supervisor who drops one of these mumbles on you? If you plan to meet it head-on by asking him to slowly repeat what he has said (he won't remember the words), by all means, ask him in private. Embarrassing the supervisor in front of his workers will backfire on you sooner or later.

If you realize that the expression was a smoke screen, a phony, then the best bet is to let it slide by without comment. If it should come up again, there is one more suggestion to address the supervisor's concern (whatever you think it is). A few days after the three magic words, you might try approaching the supervisor with an outline of what you think he may have been alluding to, and ask him if he wishes you to work on this project with him. Chances are that he will take it back to his office, and you will never hear about it again. Why bother? Because if you recognize this weakness in the supervisor but absolutely cannot face him with it, then inventing a meaning to his words and showing him a positive interest in helping him will, if nothing else, earn you some "cooperative, helpful, friendly points" that might serve you well at some time in the future.

Another thought on mumbling: this language of mumbling is an important part of the legal language, and because many politicians are also lawyers, they, too, have taken to mumbling in their regulations that may apply to your industry. Perhaps, to be tactful, you might wish to consider the use of a more acceptable word when asking for clarification from regulatory agencies: "legalese."

### Do Not Confuse Buzzwords with Mumbling!

Caution: there is a world of difference between the use of buzzwords and the use of mumbling. Nobody understands the meaning of mumbled talk, and it can successfully be ignored. Not so with buzzwords. For an indeterminate period of time, buzzwords are usually taken quite seriously and are more or less understood by the person using them. If the quality manager has any doubts about what the CEO of the company means when instructed to set up a benchmarking project on the Finnegan Food plant, he must not treat this as a mumbling event. He is compelled to find out what the CEO has in mind and to proceed with the project. If the quality manager is unsure of the meaning of "benchmarking," he might consider a tactful approach such as "I'm a little unsure of the latest techniques in benchmarking. Can you suggest where I can brush up on it—perhaps a book, or a seminar?" It is just possible that the CEO had in mind the search for an internal production standard derived from some "best practice" of the recent past. It might create an unpleasant scene if the quality manager, guessing at the meaning of "benchmarking," decided to buy samples of every Finnegan Food supplier in the country as a starting point, or, worse still, the quality manager might have decided to contact the competition and inquire about quality.

Once more, a word of caution: make sure that you recognize the difference between mumbles and buzzwords before deciding on a course of action. If in doubt, find a tactful way to ask.

Occasionally, well-structured quality systems are installed by enlightened management and are labeled with some acronym to identify the new program and to create enthusiasm. Over the years, either through management changes or shifts in company priorities, these systems degenerate into meaningless buzzwords. Quality management should be aware of this possibility and should watch for program deterioration so that they do not find themselves in the position of "kicking a dead horse." If the program has fallen into disrepute, it should not be mentioned in the course of business. Two cases have been observed where total quality management (TQM) had been carefully installed and operated for several years and then abruptly dismantled without notice.

In one instance, a large processor had constructed a TQM department that effectively cemented the quality activities of all phases of the company operations. A few years into the program, a staff quality engineer was troubled with some calculations involved in a statistical survey he was conducting and telephoned the TQM department for expert assistance. This was not an unusual request—he had received such help many times in the recent past. On this occasion, the telephone continued to ring for several minutes until answered with a "yes?" Puzzled, the engineer asked if he had reached the TQM department and was told that "I used to be the receptionist here, but the TQM department was closed down yesterday afternoon, and I'm just packing my personal stuff into a box." It's difficult to understand the management thinking behind this action, but the move might have been the result of a burst of anger, a change in personnel, or some financial or legal consideration. Management failed to take advantage of an opportunity to gracefully exit with a general announcement that "since TQM was proceeding so well throughout the corporation and each department is now capable of continuing these efforts without further support, the TQM department is no longer necessary. It will be closed on Monday the 25th."

A more brutal treatment was experienced by a major process industry company that had embraced many of the latest techniques in the fields of Quality Control, Process and Product Improvement, Total Quality Management, and others. After months of secret meetings and accompanying rumors, the company was merged into a major conglomerate. A few weeks later, the vice president of quality assurance was conducting a routine weekly brainstorming conference with several other department heads when the new CEO burst into the room and announced that "this meeting is over, and this is the last of them. Now will all of you quit wasting time talking and get back to doing your jobs!" Obviously, there was no room for discussion here. The one and only alternative was to shift gears and abandon all of the newer techniques until the wind died down.

TQM carries the same air of authority as "quality assurance" introduced in the 1950s. Both buzzwords will probably remain in the quality control vocabulary for many more years. A brief list of definitions of TQM was offered in Chapter 2, but it hardly covered the field. In an attempt to explain what TQM means to the restaurant business, R. J. Alvarez presented a paper at the Institute of Food Technologists 1993 annual meeting, which covered a large number of references on the subject. As you skip through this list, try to visualize which of your executives is capable of accomplishing the implied tasks, to say nothing about even remembering them all.

1. Continuously improving an organization through methods, concepts, and philosophy
2. A system of sustained processes for improving the quality of every facet of the company, including external and internal customers
3. Total top-down commitment for managing employees and working with customers and suppliers
4. Long-term visions and values that assist managers in making immediate, tangible improvements
5. A series of systems that sustains the drive of the entire company toward continuous improvement of goods and services
6. An endless statistical quality control process focused on customer satisfaction
7. A combination of techniques including employee empowerment,

establishing benchmarks of performance, improving company processes, satisfying customer requirements, reducing defect production with associated costs, and control of cycle times

8. Continuous application of the Deming plan/do/check/act cycle
9. A method in which management quality is the goal, rather than management of quality
10. Goals stated in the following partial quote from ASI, 1990 Total Quality Management:

A TQM organization is . . . committed to involving all employees in a cultural change to more effectively utilize the total human potential by maximizing self-actualization and eliminating fear; empowering lower levels to act with shared information; sharing rewards . . . ; integrating strategies that are balanced with new technology and culture; showing day-to-day balance in management style and decision making; being socially responsible and environmentally conscious; providing a clean, safe and enjoyable work environment; focused, actions are prioritized; and directed toward a vision by a leader.

It is a rare CEO who can understand and digest all of these concepts and then, with complete confidence, announce to his managers that "Having looked at all of the quality control systems available, we have selected TQM as our corporate quality policy."

## REWARDING SUGGESTIONS AND IMPROVED SKILLS

Well-thought-out suggestions for improved process or quality, from any level employee, should be encouraged with a suitable reward. Accompanying the reward should be some form of recognition. In addition to the basic needs of health, safety, food, clothing, and shelter, people need self-esteem, and in the workplace this is best obtained from fellow workers, supervisors, or managers as recognition. It can appear in various forms. From peers, it may be a pat on the back, an encouraging "good job, Charlie," or, where a company has a formalized program, an award by a peer group. From management, it might be a medal, a gift, or a cash award, and preferably, it should be awarded publicly.

Recognition from management should be balanced. If profit-oriented departments (such as sales) receive most of the recognition, it is probable that quality-oriented departments (such as Production and

R&D) would feel left out, with a resulting negative effect on their efforts. If teams are the winners of most recognition, then individuals may tend to consider themselves losers.

Whatever programs are used by the company should be made available to all employees in the form of a pamphlet or perhaps a video. The programs should be monitored periodically for effectiveness, preferably by a skilled outside consulting firm. Employee input should be welcomed during these evaluations. Following are a few examples of suggestion and reward plans, some winners and some losers.

At an American electronics plant, a new company policy was announced that promised pay raises for learning-improved skills. At the outset this looked like a win–win situation: the employees, through their own efforts would increase their capabilities and earn more money; the employer, with only a modest outlay of funds, would benefit from a more skilled work force. In reviewing the program after 10 months, some very serious flaws appeared, not the least of which was that there was no reasonable way to stop offering it.

Because of the pay-for-learning carrot, far more employees than expected took advantage of the offer, and the costs skyrocketed considerably over budget, starting in the second month. Because the company left open-ended the nature of the skills to be learned, it became apparent that some of the skills would become obsolete after a year or two because of product changes, but the added pay would continue. Worse still, some workers took extra training courses solely for the pay raise and did not use the acquired skills on the job.

Perhaps the most important drawback to the program was that it encouraged *individual effort,* destroying the team concepts that had been built over the previous years. The idea of *team effort* was to encourage all to participate in improving product and process, regardless of formal training and abilities. Now this principle was being destroyed.

Another reward plan was devised by a medium-sized company: the employee who submitted the best suggestion during the month would be invited to dinner at a fine restaurant with the plant manager and his wife. Probably the only good feature of this program was that it was inexpensive. The faults should be obvious: most employees would feel ill at ease having dinner with people from a different social level; the plant manager's wife had nothing to say about the plan and might resent the intrusion into her life; fellow employees might harbor jealousy or resentment at the winner's good fortune; the manager would likely tire of this commitment after a few months; and, as in the education-based reward above, there was no tactful way to end the program.

These examples are not meant to imply that suggestion plans and education programs are obstacles to quality control and improvement. Rather, they are mentioned only to emphasize the need for careful planning when constructing a system for rewarding employees' ideas. One company built one of the more fruitful plans around a balance of encouragement and fairness to originators of successful ideas. Supplies of suggestion forms were placed in highly visible wall boxes located in offices, hallways, and lunchrooms of each of its plants. The forms were multipart so that several copies could be easily prepared. The format required that the originator(s) detail the existing system, proposed changes, estimated cost, benefits, and probable time to install. Clear instructions were included regarding departments to consult for assistance in preparing these estimates. Upon completion of the suggestion form, the initiator removed one copy for his records and sent the remainder of the packet to his immediate supervisor, except for a small dated tear sheet coded with the packet's serial number. The coded sheet went directly to the manager for follow-up. The supervisor was required to forward the packet within 10 days, with his written reasons for approval or disapproval, directly to the site manager for action.

Within another 10 days, a copy of the suggestion would be returned to the originator(s) with estimated time of completion (if accepted), or reasons for rejection. Reward for successful suggestions consisted of a fixed percent of the first year's savings. The company newspaper provided generous coverage of all active suggestions in the program, with testimonials for and pictures of the originator(s).

This well-designed system may appear to be unnecessarily complicated, but, without the "complications," it probably would have lost its usefulness quickly. The 10-day requirements insured that the project would not "die en route." The fixed percentage reward eliminated any doubts about unfairness. (Some of the cash awards were very significant.) The flow of paper kept the project moving and the originator(s) informed. The recognition on completion provided encouragement to others.

Looking for suggestions from all levels of employment is not a new concept. On March 25, 1949, the subject of an unusual interoffice memo from the manufacturing division vice president of a major New York corporation to the director of quality control was "Ten Commandments for Technical Men." One of these read, "Thou shalt seek and respect the opinions of operators, even unto the third helpers, for theirs is a wisdom unknown to technocrats."

Perhaps the oldest suggestion system on record was established in

1932 by McCormick & Co. It created a remarkably successful "junior board" for new ideas. This is discussed at some length in Chapter 5, along with some more recent highly formalized techniques of generating quality and process improvements through the "team concept."

Having explored several suggestion plans, the question arises, "What has this got to do with choosing a quality control system?". Some suggestions that contribute to quality or process improvement originate in areas other than the quality control department. The front office receptionist or the plant electrician need not be employed in the shipping department to have a useful suggestion to alleviate truck traffic in front of the factory. The more even-handed, open treatment of employee suggestions, the better the morale. The more people in an organization who contribute quality and process improvement ideas, the better for all. Quality is, indeed, the responsibility of the quality control department, but it is not their sole property.

## INTRODUCING NEW QUALITY CONTROL TECHNIQUES

It has been said that, in ancient Rome, whenever an engineer designed a structure, he was ordered to stand beneath the last completed arch while the supporting scaffold was removed. This clearly drove home the need for integrity in engineering quality! There is a lesson to be learned here. Chances are that the quality manager won't be required to actually operate a production process that uses one of his newly designed quality techniques, but wouldn't it inspire confidence in his associates if he made it a point to be on the line the day it started up, actually taking quality data, plotting charts, and looking really concerned?

One of the recent popular quality control techniques that has received considerable press is the so-called "six-sigma" system (Figure 3.3). It derives its name from the fact that, if the specification limit is beyond six standard deviations of the mean on both the high and low side, in the long run, there will be no more than 0.001 defects per million units produced.

A practical modification to this stringent control plan is known as the 3.4 ppm (parts per million) or 3.4 defects per million program (Figure 3.4). This plan is a compromise between near perfection and practicality. It consists of setting the specification limits at six-sigma on either side of the mean and then shifting the process mean by 1.5-sigma toward the upper specification. Without going through the mathemat-

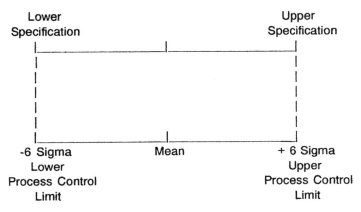

FIGURE 3.3  Six-sigma plan.

ics involved, the net result of this system is that there will be zero defects on the low side of the specification and no more than 3.4 defects per million on the upper specification limit. This procedure is highly successful where the lower specification is supercritical, and 3.4 defects per million on the high side is acceptable to the consumer.

Certainly, this is a most desirable state of affairs. The customer should be delighted to purchase a defect-free product. Several questions arise: Is this level of quality really needed? Can the manufacturer afford to provide the product at this level of perfection? Is it possible to reach this goal? How much higher can the manufacturer raise his price and still keep his customer? If the product is a newspaper, how important is it that there be absolutely no misspellings in an edition? If the product is a bottle of 1,000 saccharin tablets, how serious a problem is it if more than 3.4 tablets in 1,000 bottles are chipped?

There is no question that, if the product or service is critical, six-

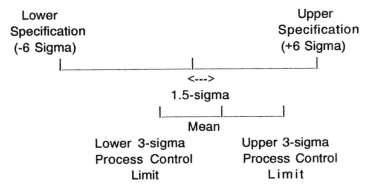

FIGURE 3.4  Parts per million plan.

sigma specifications may be called for. Some applications really deserve this level of quality: foods, drugs, medical services, aerospace, and some electronic and finance applications come to mind. Furthermore, many times this goal is economically obtainable, but, in the past, the company had not considered it necessary because of lack of competition or ready acceptance by the customer. A half dozen loose threads on mechanics' wiping rags in every shipment would hardly be considered a defect, but a split thread on a surgeon's suture material at any time would be a disaster.

On the other hand, six-sigma quality control is an excellent goal, and there are several techniques available to reach it:

1. Expand the specifications to 6-sigma
2. Tighten the control limits to 1/6 sigma
3. Combination of 1 and 2

Example: A company manufactures syrup pumps that are claimed to move 10 gallons per minute. Tests were run on several hundred pumps, and the standard deviation of delivery was found to be 0.1 gallon per minute. Six standard deviations would permit a delivery of 10 ± 0.6 gallons per minute, or 9.4 to 10.6 gallons per minute, and this could be guaranteed for all but 3.4 out of a million pumps sold. (This example has been simplified. To be absolutely precise, the process center would be shifted 1.5 standard deviations above target to 10.15 gallons per minute.)

### Problem Solved by Technique 1

A potential customer wishes to order several pumps specified to deliver between 9.5 and 10.5 gallons per minute. The pump manufacturer cannot quite meet this goal. If the customer can be persuaded to stretch his requirements by 0.1 gallons per minute on each end of the specification, nothing need be changed. Persuasion might have to include a price reduction, but it is possible that the customer had dreamed up his specification out of thin air and would be perfectly satisfied with the 9.4 to 10.6 limits available.

### Problem Solved by Technique 2

With some brainstorming on the part of the workers, it may be possible to make some minor modification in the process to meet the 9.5 to 10.5 requirements. Perhaps a little extra care in polishing off burrs in the cast housing or the discharge orifice or making some other minor change in the process would suffice to reduce the standard deviation. If these modifications were found to be impractical, it might then be necessary to have engineering redesign the pump, perhaps replacing the cast impeller with a precision molded unit, to meet the stringent delivery requirements.

### Problem Solved by Technique 3

Work with the customer to expand his specification needs "a little bit." At the same time, work with production and engineering people to make slight improvements on the delivery standard deviation. Possibly a compromise could be reached at, say, 9.45 to 10.55 gallons per minute.

Consider this more realistic example: operating at 3 sigma limits, the manufacturer would produce no more than 15 units per 10,000 pumps beyond either the upper or lower capacity limit. Probably, the upper limit is not critical to the customer. If 10,000 units represents 5 years of sales, however, then the management is probably going to think twice about spending time, effort, and money improving this seemingly adequate performance. He can live comfortably with three defects per year.

Before attempting to propose "a really neat new quality control program" to management, it is essential that the cost, benefits, and practicality be clearly thought out.

### Selling the Electric Light to Candle-Burning Management

#### Ask for Help

Introducing a new quality technique to a management that rarely takes to new ideas calls for a careful strategy and considerable tact. For example, a coffee roaster company purchases raw beans from several suppliers. These are subsequently stored, blended, roasted, and ground. Moisture content of the raw material is important because it affects the process yield and cost and has been determined for years by an "old-timer" purchasing manager who claimed he could squeeze the beans in his hand and classify the moisture level to the nearest percent. One day, the shiny new quality control supervisor approached the old-timer with the suggestion that the lab could provide accurate and reproducible moisture content of incoming samples for his guidance and presented him with a tray of beans as an example. The old-timer squeezed a handful from the tray and pronounced the moisture as 12.3%; the quality control supervisor then told him that a moisture testing device found the level to be 12.5% and complimented him on his unusual gift of ascertaining moisture.

Obviously, this approach achieved nothing. Quality Control should have known better than to present a bolt-out-of-the-blue test to the old-timer. Suppose the old-timer had pronounced the level to be 11.0%. What then? Perhaps an argument over who was correct? Wouldn't there have been a far better chance of cooperation if the quality control supervisor had initially approached the old-timer with a hat-in-hand request for his assistance? Make up any reasonable story. How about this one: "Good morning, old-timer. I really would appreciate your help on a problem the boss just dropped in my lap. He wants to insure that the moisture level of the raw material is within specification on those days you are either out in the field or on vacation. Could you help us by checking our moisture analysis method for a few days and comparing our results with what you know is the correct value?" Chances are that, within a week or two of this cooperative effort, the old-timer will suggest that, from now on, the lab should run all of the determinations and should send him a copy of the results.

## Plant and Nurture Ideas

Another situation: the multi-head bag filler operator returns from a three-day seminar on quality control techniques, which included an emphasis on attribute control charts. This is a totally new concept to the press operator. He is burning up with the excitement of knowing that his supervisor has been using the wrong quality control technique all these years and can hardly wait to tell him. This head-on approach could be a disaster for the operator. The only tactful way to present the information would be over lunch, where he discusses *all* of the high points of the seminar. If the supervisor doesn't recognize his need for attribute chart controls, for example, the operator should be careful not to force the subject. As an alternative, he might show him applicable class notes and perhaps emphasize the concept with a couple of examples until the supervisor eventually chooses to try out the idea. Some theoreticians in the field teach that the boss will readily embrace any worthwhile new quality control idea from an employee, but the real world doesn't work that way. Too often, the boss tends to feel embarrassed, threatened, or even angry when good quality control ideas bubble up from the ranks below. It is always better to plant and nurture ideas, thus allowing credit to go to the boss.

## Sell Ideas by Little Steps

When attempting to present a new idea, design the ideas as if they were a sales campaign. Most new ideas demand it. For one thing, everyone is busy and would rather not have their routine changed. Rarely does one have the chance to say, "This is a new idea and it will be implemented as follows." It will find better reception if introduced with, "We've been working on a new idea. What do you think of this?" Plan on taking little steps during the project presentation. Start with a sure thing—an idea that will be readily accepted, even though it is not critical to the overall proposal. Then, pause. Build up a small amount of suspense. If successful with this strategy, the "customer" will look you up eventually and ask what the next step is.

## Create a Partner by Asking for Help

It is generally useful to ask for help along the way, whether needed or not. When someone lends a hand to a product or process improvement program, they become partners, even coauthors. As a successful program unfolds, they will be somewhat eager to take part of the credit and will do what they can to assist.

During the study phase, it sometimes helps to make small errors and allow the supervisor to correct them. On the other hand, during the implementation phase, mistakes are the last thing one needs.

## Find and Advertise the Dollar Value

Whenever possible, a product or process improvement suggestion should include costs, savings, or both. The dollar is a language that all can understand, whereas sigma, control charts, or acceptable quality limit may be misinterpreted by others. Where costs are not readily available ("that's confidential information" or "I'm pretty busy with the corporate balance sheets, come back next month") make the best estimates possible. If factory prices are not forthcoming, use the retail price—someone from the accounting department will come up with the correct figures. They probably know that someone in upper management will eventually demand them.

## Avoid Looking for Credit

As a rule of thumb, it is wise to appear with the highest profile in dull routine quality activities. Conversely, maintain low profile in major quality projects, allowing others to enjoy the glow of major successes. In the first place, the world around you will understand where the project originated; the credit will arrive eventually. In the second place, those enjoying the fruits of your labors will be willing partners to future efforts.

## Advertise Quality at Every Turn

Most other departments manage to keep themselves pretty busy without going out of their way to get involved with quality problems. To be truly effective in developing and maintaining a company-wide quality management effort, it is necessary to lobby everybody with discussions about quality subjects. Talk about your projects with everybody—frequently. Every so often they will come up with some useful quality suggestions you might not have considered.

## Watch Your Language!

Avoid technical jargon and buzzwords when talking to a nontechnical person. A statement such as "all that is required to narrow the three-sigma limits and simultaneously improve the attribute charts is to . . ." will probably be all the listener will register. The rest of the sentence will likely be lost while he tries to remember what "three-sigma limits" means. If it's necessary to use technical expressions, take the time to add one more sentence to explain what they mean. With the common usage of computer charting, a production line operator merely presses a key to see a control chart, complete with upper and lower control limits, but, after a while, he is likely to forget what the technical term is for those limits. Talk slowly—nobody really understands your job as well as you do. Avoid embarrassing your listener.

## Sometimes the Government Is Your Friend

Sometimes, it helps to quote government regulations to sell a quality control procedure. Compare the effect of these two statements:

1. If we purchase this $5,000 extraction column, the lab will be able to control the cadmium content better.
2. The government regulation XYZ123 limits the amount of cadmium we are allowed to dump in the sewers. We can control this quite simply by adding a $5,000 extraction column to our high-performance liquid chromatographic (HPLC) meter.

## Bring Potential Allies to Seminars

How can you sell quality control to some other department? Supposing you have just learned about a new technique for controlling quality of sales. Attempting to sell it to the sales department might well be considered an intrusion on your part. If, on the other hand, you were to ask your supervisor for permission to attend a seminar on quality in sales, and could provide sufficient background information to make it seem worthwhile, this might be a route to possible support by management at least to explore the subject. If a representative of the sales department were to accompany you to this seminar, it might assure cooperation from that department. Even if the request is turned down, at least it demonstrates your interest in improving your skills.

## Try Mind Reading

Put yourself in the shoes of the decision maker when promoting a project. Look for signs that the boss is in the right frame of mind to consider your suggestions. If he has a hat full of problems at the moment, he is likely to respond, "You're right, but forget it; I'll take the responsibility if anything goes wrong." In the event something does go wrong, chances are that the boss will have forgotten he ever said that. Wait for a quieter moment to bring him new ideas. Had the same project been proposed when the decision maker was more at peace with the world, his reply might well have been, "You're right. How can I assist you in getting this project started?"

Another statement from a frantic supervisor: "Don't worry about cost; I can hide that. Production is king. Get it into the warehouse."

This probably means that the supervisor has just been read the riot act by a superior. The difference here is that he probably does mean what he says—he *can* hide the cost. By all means, help him; some day he will return the favor.

### Choose Proposal Vocabulary with Care

Sometimes, the ideas are fine, but the words describing them are unacceptable. Take, for example, the concept of adding water to a granular food product to maintain the highest acceptable level that will affect neither quality nor shelf life. Suppose that a sequential chart of moisture values shows that there are natural seasonal cycles. If the upper level of moisture is found to provide a satisfactory product, calculate the dollar profit that might accrue from artificially maintaining the moisture at that level over the entire year. Suggesting to upper management that water be added to the product, or that the moisture level be adjusted, or that the water content be controlled, or that the end product be moisturized might all be rejected as compromising the reputation of the company in the event the word was spread throughout the industry. On the other hand, the suggestion that the end product be "tempered" or "quenched" or "humidified" might more likely be embraced as a suitable description.

### Set Quality Goals

When a quality manager attempts to set his goals, he usually finds that company quality policy is nebulous. "We are going to be the best in the business, and we expect to constantly improve our quality and productivity." In reality, there is a quality level beyond which the company will no longer be profitable, but this can never be written into the quality manual. It is usually fruitless to abruptly confront upper management with a request for their concept of "product quality." They generally will answer in meaningless superlatives such as *the finest in the industry, world-class, top-flight, upper-crust, untouchable, customer delight,* etc. One foolhardy quality manager suggested that quality might be defined as "goodness," a term that the boss quickly pointed out was sharply debated by every religion around the world. In attempting to define a meaningful company policy on quality level for inclusion in the company quality manual, it may be politic for the quality manager to meekly send a proposal to the upper levels of management, using relatively bland terms such as "competitive," "highest level commensurate with our profit policies," or "equal to or superior to other market leaders." Invariably, the proposal will be returned with the modest term replaced with a slightly stronger one. That will probably suffice until the next revision of the quality manual.

## Crisis Management by Edict

The boss is furious—50 cases of product have been declared off-sale for short weight in Pennsylvania and, the very next day, 30 more cases in New Jersey. This comes on top of a consumer complaint of an "empty can." Executive statement by the CEO to the quality control manager: "From now on, every can shall be exactly 16 ounces. There shall be no more underweights."

There are many ways to respond to this:

1. Redefine "underweights" and expand specifications.
2. Inspect 100% and remove underweights.
3. Capability study—can process meet existing specs?
4. Control chart analysis of a testing program for process improvement.
5. Purchase a more reliable weighing device.

When the smoke clears and the noise dies down, quality control might explain that the plant has probably *never* produced a can with exactly 16.000000000 ounces. How much variation should be allowed? 0.001 ounce? 1/16th ounce? If this has not yet been established through statistical testing, now is certainly an appropriate time to start, and presents an opportunity to promote statistical quality control techniques to the CEO.

To establish present weight variation, should one use gross weight or net weight? How much does the tare vary from container to container? What is the weight loss due to packing under vacuum—air does have weight. How much weight is lost from opening a vacuum-packed can caused by carbon dioxide escaping from the product—it weighs even more than air. How many cans should be selected at what interval for this information? There are many causes for this variability, and they must be identified if they are to be eliminated. Because a serious weight control problem seems to exist, it seems logical to suspect that an adequate quality control system has not been developed in this company. Following is a suggested approach to finding a solution, assuming that the product is processed and canned in-house.

1. What can the filling scale do? With product running with perfectly uniform moisture, carbon dioxide, temperature, pH, density, etc. weigh 100 cans from each of the plant's scales, and establish standard deviation of each scale. Calculate sigma of the scale system.

2. If this sigma is satisfactory to the boss (and it rarely is), establish control limits for a control chart. If not satisfactory, investigate the causes of variability for each variable. Carbon dioxide, for example, may vary from changes in a mixing hopper, age, temperature of process, variety of raw material, height of product in feed bin, speed of screw conveyors, etc. Then proceed to the next variable: moisture, for example.

3. Brainstorm possible methods to limit each variable. Use plant operating employees as well as technical staff for the brainstorm sessions. Guide the brainstorming effort by constructing a process flowchart, starting at the right-hand end (the filled container at the shipping dock) and working backward. Be sure to include all contributing tracks. Chances are the flowchart will consume an entire blackboard upon completion. Figure 3.5 is an example of the first attempt at constructing a flowchart. (See also Chapter 5 discussion on charting.) This seems too simple a tool to be of much use. Looking at each step more carefully reveals that several steps have been left out. An expanded chart should then be prepared (Figure 3.6).

4. Select the most likely areas of variability in the process flowchart and construct a cause-and-effect diagram (fishbone chart) of each, showing possible areas of interaction. In this case, perhaps someone has observed that many times the decolorizing process changes the product flow and density, thus possibly affecting the weighing process. A fishbone chart (such as Figure 3.7), might help in isolating critical points in the process.

In this example, it may have been observed on several occasions in the past that vibration or carbon dioxide have been likely causes for producing unusual densities in the decolorized product, subsequently resulting in poor weight control. Having pinpointed these causes, various analytical procedures, such as those discussed in Chapter 5, may efficiently produce a solution to the problem.

One more very important point: unless otherwise directed, it would be wise to keep the CEO informed of every step of this program. By letting him know of the powerful quality control tools available, he

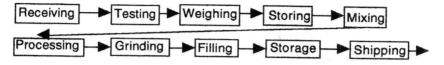

**FIGURE 3.5** Process flowchart—first attempt.

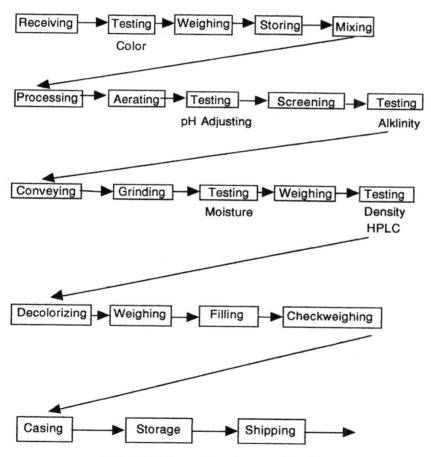

**FIGURE 3.6** Process flowchart—revision #1.

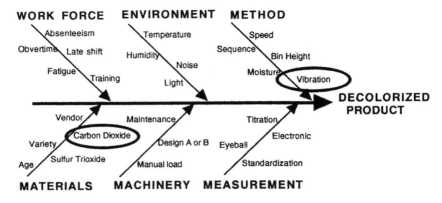

**FIGURE 3.7** Fishbone analysis.

will be in a better position to establish a realistic and useful company quality philosophy. As a side benefit, he will be less likely to make emotional outbursts, such as this one about net weights, and should become more receptive to further quality control efforts.

### While You're Away, Attending Classes

If you are a member of a quality investigative team and are called away from your job on the production floor to attend a seminar or a class, it is necessary to stay current with the team's progress by checking with a team member at the end of each day. This is important, not only for "job insurance," but to stay current with the team's accomplishments so that, upon return, you are able to provide immediate input without spending the next few sessions trying to catch up. Production personnel are of special value to team operations because they are most familiar with the hour-to-hour procedures and problems.

If you are the team spokesperson or are responsible for the implementation of a policy or procedure that the team has proposed, it is necessary that you call everybody involved daily. Don't let them forget you or change things. If a problem in implementation arises while you are out of town, never plan to solve it "when I get back." Arrive at some kind of resolution over the phone, and immediately report any kind of news, favorable or unfavorable, to the boss. Any break in the constant communication flow between quality control and production supervision may result in one of those nonrational decisions to "ship it just this once—I know the quality's not right, but maybe nobody will notice."

This discussion on "selling the light bulb" may have seemed a bit elementary—and it is. But the principles of promoting solid statistical and quality engineering techniques both up and down the organization ladder, and maintaining close relationships with all, are an integral part of a successful quality control system and are far too often overlooked.

## CONTINUOUS QUALITY IMPROVEMENT

A quotation from a *Readers Digest* article by David T. Kearns (May 1990) says: "In the race for quality, there is no finish line."

Continuous improvement makes much more sense than attempting to devise "the one perfect solution." Understand at the outset, however, that quality improvement is never good enough—only perfection will do. Expect complaints about improvements or partial solutions. Here are a few examples:

1. You're a little kid. You decide to improve the appearance of your room by putting things away. In the process, your box of dried grasshoppers spills out all over your bedroom floor. You spend 15 minutes laboriously picking them up and returning them to your box, which makes you late for dinner. Mother asks, "Can't you ever get to dinner on time?" One hour later as she tucks you into bed—crunch!—she steps on the one grasshopper you missed. "Why can't you ever pick things up around here? If this happens again, there'll be no television for a week."

2. A major automobile manufacturer decided to reduce the number of bumper defects. For the past few models, the number of chrome-plated bumpers that were blemished in one spot or another amounted to approximately 19% of those produced. For the most part, these imperfections were considered a normal part of the plating operation, and rarely were any units actually rejected. An engineering study found that condensation in the discharge ventilators over the plating baths was dripping back into the plating tank, producing these defects. A redesign of the ventilators reduced the blemished bumper rate to 14%. A second and third study found other causes. After corrective

action was taken, the defect rate had been reduced to 4%. At this level, management decided that the problem had been solved. But, 4% of the new car buyers may not have been equally pleased to find blemishes on a conspicuous part of their purchase.

3. When should continuous improvement finally stop? Is planned obsolescence an unsatisfactory goal? Must continuous improvement and "customer delight" be a never-ending quest? Approximately 50 years ago, an appliance manufacturer developed an excellent electric razor and, over the next two or three years, made continuous quality improvements. Now, 50 years later, the razors still work, but the electric cords are nearing the end of their useful life. One cannot buy either a cord or a complete replacement—the manufacturer went out of this business 20 years ago because the market dried up. Of course it did—the product had been improved until it was "too good."

4. One Saturday afternoon, a homeowner handyman went to the local hardware store and purchased a garbage disposal assembly for his kitchen sink. After spending the afternoon installing it, he found the unit to be jammed. He disassembled the unit and brought it back to the store the following Monday for an exchange. To make his work simpler, he left the upper four-lug assembly unit attached to the sink, returning the rest. Back again under the sink, he found that the replacement unit was a "new and improved" model with a three-lug attachment head, which obviously would not fit. This may have been considered an improvement by some, but not to this frustrated customer. Perhaps any "new and improved" product idea should be discussed with the marketing department before final implementation.

## New or Improved Product Introduction

Research and Development has designed a new product; Accounting has worked out the cost and profit projections; Human Resources has arranged for training and staffing; Distribution has contracted for delivery and storage facilities; Engineering has laid out the process lines and the packaging; Production has determined procedures and schedules; Marketing has arranged for advertising and prepared the sales force; and the CEO proclaims to all that you've all worked well together as a team, and this product will be a real winner.

Enter the quality control manager, "We have prepared quality safeguards and quality participation along the entire system; here are seven critical areas in which this product might fail." The CEO responds: "The trouble with you is negative thinking. Be a team member; think positively." (There is an implied "or else—" in that statement!)

How can Quality Control perform its primary function of preventing problems and improving quality without looking at the things that might possibly go wrong? Perhaps the CEO is correct: the quality control manager's statement is at least a partially negative approach to the project. Would it not have been more effective and more acceptable to present possible improvements and solutions that would avert the potential failures?

## Customer-Driven Continuous Quality Improvement

In most cases, process and quality improvement is based purely on the manufacturer's search for lower production costs and higher profitability. Maybe it is customer driven, but, in some cases, a quality change based on the customer's desire for an improvement is not enough to generate a market; the manufacturer may have to advertise heavily to convince the consumer that the improvement asked for is now available. For an example, look at the history of the green pea.

There are references to the growing of peas approximately 4,000 years ago in the upper Nile Valley. Reason would tell us that dried peas were discovered in the same era, since those pods left on the vines beyond harvest time would lose moisture in the hot sun. Therefore, it is probable that those early inhabitants' desire to have peas available during the winter season was easily met by the growers of those times. On the other hand, sun-dried peas did not measure up to the quality of the fresh product. They had a different flavor and often were moldy, contaminated with rocks, impossible to eat raw, took too long to soak and cook, and had an unpleasant gray color.

Improvements were slow to come. It took until the 1600s before the quality of dried peas was improved by inspection. With the removal of some of the defects, the dried peas gained acceptance for soups, stews, and mushlike recipes. This product was relatively stable over the postharvest seasons and was part of the standard fare for lengthy excursions of ships and armies. It is expected that this customer-driven quality improvement was accepted without a major sales effort.

The preservation of food in airtight containers was demonstrated in

1745 by Needham, an English scientist. Several years later (1804), Nicholas Appert developed the first successful glass jar pack of foods and shortly thereafter founded the "House of Appert" food packing company. By 1810, Peter Durand patented the process of preserving food in metal cans, and the canning industry took off. Finally, the customer demand for off-season, easily prepared, nearly natural-textured vegetables was met.

The next series of quality improvements was more likely the result of profit greed on the part of the manufacturers than customer needs. The first cans were cut by hand from sheet, rolled, and individually soldered. Later, the ends were turned up by a mallet and a "heading stake" to facilitate soldering. Twenty years later, the vent hole system of heating and sealing was devised, simplifying the final solder operation. An expert tinker could produce up to 60 cans per day, and these were made at the cannery, since no can manufacturers existed per se. These improvements had little or no effect on the consumer nor were they the result of consumer demand.

During the 1850s, further process improvements came quickly: a stamping machine, the pressed top can, and the floating process for soldering. As a result of these steps, one man could seam 1,200 cans per day. In 1893, the Karge rubber-gasketed and roller-crimped can was introduced to the United States from England, eliminating the soldering operations. In 1903, the Max Ams Machine Co. designed and manufactured double-seam can-making machinery, and the canning industry grew geometrically. At the time the double-seam can appeared, the canned pea industry was producing about 4 million cases per year. Fifty years later, the yearly pack ranged between 30 and 40 million cases. Here, too, it is extremely unlikely that the consumer had any part in demanding these improvements. The improvements were based on technical abilities and the search for greater profits.

Other product quality improvements were the result of technological breakthroughs:

- reduction of sterilization times by use of the calcium chloride bath, resulting in better color, flavor, and texture of the peas
- pressure retorting to further reduce the time of heat exposure
- continuous agitating pressure sterilizers

So far, it appears that most of the quality and process improvements were the result of independent efforts of manufacturers rather than coercion on the part of consumers, but the picture changed in the early 1900s. Consumers complained that the canned products were not the same as fresh peas. They had a tinny flavor, a bleached or gray color, were mushy, the size was nonuniform, some were split and the skins floated in the brine, and there were pieces of pod, vine, and rock in some of the cans. Here were opportunities for the canner to beat out his competition with product quality improvement demanded by the consumer.

The canners responded with improved plant breeding, mechanical harvesting and vining, better internal can coatings, gentler handling equipment, automatic product inspection, improved brine tank grading methods, and new sizing controls.

Sometime in the late 1930s, the consumers began to show more sophistication and have since been much harder to please. For example, the consumer complaints about dried peas were answered by the industry with the development of dehydrated (and, later, freeze-dried, precooked) product. The processors touted these products as possessing more uniform color, size, and flavor with improved keeping qualities. The consumer resisted because of perceived problems with price, availability, and cooking.

When the consumer complained about the color and flavor of canned peas, the industry responded with frozen peas. The consumer greeted this new development with a series of complaints: the color is too bright, the flavor is artificial, the cooked frozen peas are too mushy, and the product deteriorates when stored in the home. Subsequent development of this product is, perhaps, a classic example of consumer-driven quality improvement. The invention of the plate freezer provided a major improvement in overall quality; the invention of individually quick-frozen pea equipment (belt, blast, immersion, and drum freezers) and the development of new systems for storage and distribution of frozen foods eliminated most other quality complaints. Finally, an education program aimed at the consumer provided the general acceptance of this method of food preservation. At last, peas of excellent flavor, texture, color, and cooking qualities were available to the consumer year-round!

But there is still resistance by the public to newer developments. Quick-cook, dehydrated peas referred to above are still not generally accepted for home use, even though this product is flavorful, easily prepared, and shelf stable without special environmental precautions.

Perhaps the price is the problem or perhaps the frozen product is perfectly satisfactory.

Irradiated fresh peas exhibit a greatly increased shelf life without flavor deterioration. This process has been offered in response to the consumer pressure for a longer fresh vegetable season; yet, the resistance by a skeptical public has prevented the market from developing. They need to be convinced that irradiation can kill spoilage microorganisms without harming humans, and they need to be taught that spoilage enzyme systems can be destroyed without producing mysterious harmful "chemicals." This is yet another example of a consumer-driven quality improvement that has been requested, developed, offered, and not accepted.

Finally, the thought of gene-spliced green peas is viewed with apprehension by the general public. Although many fruits and vegetables that are now part of the public's daily food basket were developed by Luther Burbank through cross-breeding (the word "gene" hadn't been invented yet), they do not recognize the comparison. As with the irradiation technique, this answer to their request for extended shelf life may take a few years of education before acceptance.

To summarize, although a consumer request for product quality improvement is often the source of new product development for foods, the acceptance of the "improvement" is not always assured. The request should be considered carefully before changing an already successful product.

There is an area where consumer-driven quality "wants" are difficult to fully evaluate, sometimes contradictory, and occasionally impossible to achieve. Consider some of the consumers' wishes for improved food products:

- healthy, but good-tasting
- healthy, good-tasting, but inexpensive
- fat-free, but same tactile quality
- less packaging, but equally convenient
- less packaging with same protection
- reduced price, but identical quality
- "organically" farmed with equal appearance
- "organically" farmed with same price

Perhaps some of these "wants" are attainable, but there is no guarantee that the consumer will buy the new items, such as was the case for some of the green pea products above.

# SUMMARY

Implementing a quality control program should be fairly straightforward: find a competent quality manager; study the process, product, and service; assemble the program, and publish it. Would that it were so simple!

A quality program cannot be delivered overnight, over a month, or even over a couple of years. It should be developed carefully in little steps and offered only after the effect of each step on the financial report, on the company people involved, the customer's needs, government regulations, and the long-term marketing requirements are satisfactorily met.

A successful quality system cannot be flexible. On the other hand, it must be subject to constant review and improvement as conditions change.

# CHAPTER 4

# Quality and Process Solutions: Engineering Method

LORD Kelvin (1889) is credited with the observation, "When you can measure what you are speaking about, and express it in numbers, you know something about it, but when you cannot express it in numbers, your knowledge is of a meager and unsatisfactory kind."

In an early lecture, W. Edwards Deming stated that intuition and the eyeball are always wrong. In his book, *Out of the Crisis,* he wrote that "experience can be cataloged and put to use rationally only by application of statistical theory." The implication in both of these statements is that unless a problem can be expressed in numbers, it cannot be fully understood or optimally solved. Maybe so, much of the time. Some purists have gone so far as to state that unless you know all of the statistical data about a system, there is no way to be sure that a problem exists.

On the other hand, there are many factors in the industrial world that challenge the universality of these statements. As one lawyer put it, "I don't have to know how many rooms are in the house to see that it is on fire." Some problems really can be solved by intuition: would more light on the production floor reduce the number of mistakes made on line? Would a lubricant reduce bottle breakage on the conveyor? Would Alice cause less friction in this department than Pete? Additionally, perhaps human factors such as petty jealousy, "turf protection," or insecurity can add illogical facets to industrial quality problems that do not lend themselves to hard-boiled statistics.

In choosing a quality control system, the employee selected for this function will normally consider the seven tools of quality control as the primary structure, and this is certainly a logical beginning. Added to these would be governmental constraints and industry standards, and these are essential. However, often overlooked until an emergency arises is the planning for two other major types of quality problem solution techniques that are based on specialized knowledge acquired at

**FIGURE 4.1** Essentials for quality control and quality improvement.

schools or on the job (Figure 4.1). One of these is based on engineering or intuition and will be discussed in this chapter. The other is the team approach, in which several people with differing backgrounds and interests develop solutions that might not be apparent to an individual. The team approach will be considered in the following chapter.

Once these three major blocks of a quality system are in place, additional concepts such as total quality management, reengineering, benchmarking, quality function deployment, design of experiments, breakthrough, transformation, empowerment, and other quality control ideas of the day can be safely investigated without upsetting the essential functioning of the continuous control and improvement system.

There are borderline cases where intuition works well, but statistics might work better. For example, the president and founder of a fast-food chain was approached by a computer systems analyst who offered to set up and operate a program to search for optimum locations for new restaurants. The president responded with "listen carefully, young fellow. I grew this chain of 800 restaurants from a single little store, and selected each of the business sites by myself. I used my own intuition, based on 21 years of experience, never used a computer program, and made the best choices in all but two."

There is little doubt that intuition worked well for this person, but perhaps the use of statistical analysis might have done even better. It is possible that a mathematical model may have selected 800 winners, not 798. It is also possible that the profit margins might have been significantly improved by the use of a statistical selection.

The engineering approach using training, experience, and intuition as tools can often be very effective in solving quality problems without the use of statistical tools or the need for a formalized procedure. On the other hand, building a procedural framework based on "how, what, when, where, why, and who" might speed up the process, particularly if an inexperienced engineer or team has been charged with finding a solution. This technique has been used successfully for years as a guide to newspaper reporters in gathering information for a story. There are many variations of this structural procedure; some additional pieces of the framework might be "under what conditions, how often, which source."

Each industry needs to define the steps in the framework to fit its own special needs. In the hotel industry, *who* might refer to the desk clerk, the guest, or the laundry staff; for a processing industry, *who* might refer to the customer, the line operator, the shipping clerk, or the receiving inspection department. These definitions can be as simple or as complicated as necessary for the type of problem and experience of the engineer or team.

"What" is the most important step, either with or without a formalized framework. It requires an accurate definition of the problem to be solved. Before embarking on an investigation, there should be full agreement to a description of the basic problem, its severity, and secondary effects directly connected to it. For example, a customer complains of burrs on deep-fat fryers, sometimes accompanied by a fractured handle and off-center fry basket. It would be inappropriate to lump together all fryer complaints as a single problem, since defects such as leaks, warped base, missing coating, or split rivets would probably be totally unrelated to the handle complaints.

"Where" refers to location of the problem:

- at the riverfront plant
- on the mitered edge of the Finnegan Pin taper
- Midwestern area customers
- near the transfer stations between lines C and G
- in the warehouse stacks, bottom layer
- under the short flaps of the shipping case
- behind the product label
- at the finish-coat painting station

The other procedural framework headings should be sufficiently obvious to need no explanation.

One of the suspected drawbacks of intuitive problem solving is the gnawing doubt that perhaps the solution found was not the best one. When confronted with a problem, one should strive to find the perfect solution, but not insist on it. If the first solution uncovered is a good one but it is not implemented because there is the possibility that a better one might be found with a little more study, the problem might remain forever unresolved, or, as someone once worded it "perfect" is the enemy of "good." Samuel Florman suggested in an essay on "The Human Engineer" that "engineering solutions have had a way of appearing on the scene when needed—often imperfect and carrying within the seeds of new difficulties—but solutions, nonetheless."

## ENGINEERING AND INTUITION

Following are several examples of quality problems that were solved through intuitive reasoning. In each case, the exclusive use of statistical analysis would have been of doubtful value. Some of the problems were solved through team efforts, and some with the help of outside consultants, and some by individual talented or motivated employees. In each case, engineering, intuition, and experience provided the necessary tools. Use of statistics was not required—nor would it have been of much help.

### Departmental Conflict

In the manufacture of frozen food specialty assemblies, Department S sliced a frozen food slab into rectangular portions. Department D stamped out plastic dish compartments. Department A assembled these two components by pressing the frozen block into the dish. When the edge of the block did not perfectly match that of the dish, the pieces could not be combined, and the scrapped pieces were sent to Department R for rework.

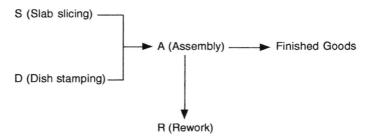

S blamed the problem on defective slicers. H blamed the problem on inadequately trained assemblers. R was baffled by the fact that some scrapped slices that were cut to specification should have fit, and some of the reworked scrapped slices didn't fit. A company engineer studied each department's procedures. He discovered that Department S and Department R were using two different gaging techniques that measured the slice curvature from different base points. Once this was corrected, the defect rate dropped to zero.

## Hydraulic Coupling Failure

A manufacturer had been successfully producing one-inch couplings for use in high pressure hydraulic systems. To assure the high quality level required for these fittings, relatively large samples were tested on a flexing device each hour of production. Over the years, none had failed the flex test. In an attempt to expand business, a two-inch coupling was designed, a prototype was built, and the testing device modified to accept the new size. The first prototype failed immediately, and subsequent samples proved no better. Several design modifications were attempted using heavier construction, larger seals, and other ideas, but none seemed to work. An engineer consultant was brought in to discover the design flaw problem. As a first step, he redefined the problem: discover the cause for failure. Using this broader definition, the testing device was included as a possible contributor to the problem, and indeed it was. By merely modifying the one-inch tester to accept a two-inch fitting, the larger fitting had been exposed to an abrupt wrenching action during the test. The tester was subsequently redesigned by scaling up the entire unit. This provided the same type of stresses as on the smaller size, and subsequent two-inch fittings were found to be satisfactory.

## Can Leakage Project

Here is an example of an arbitrary quality decision, based on the use of intuition *and* some simple mathematics by a production manager regarding can leakage. Unfortunately, the dollar savings indicated by the calculations overshadowed the serious potential quality disaster that the manager failed to observe. Intuitive reasoning would have shown at the outset that the process should not be modified.

The diagram (Figure 4.2) shows three production lines running 120 cans per minute each, with a total output of 21,600 cans per hour.

$$3 \times 120 = 360 \text{ cans/minute}$$

$$360 \text{ cans/minute} \times 60 \text{ minutes} = 21,600 \text{ cans per hour}$$

When running normally, a leakage rate of three cans maximum per 1,000 produced had been considered acceptable. Under the worst conditions, then, it is expected that as many as 64.8 leakers per hour could occur.

21,600 cans per hour × 0.003 = 64.8 leakers per hour

This rarely happens, since operating procedures state that, if two cans from the same line are found to be leaking from the same imperfection, that line is to be immediately shut down for investigation. Because of lead time through the storage loop, it is possible that more than two defective cans might be removed at the tester in the same hour. In any event, if more than 65 leakers occur in an hour, the entire plant is to be shut down, and a major investigation is undertaken until the cause of the problem has been corrected.

This procedure had been working satisfactorily for years, with virtually all leakers removed from the tester before entering the warehouse. Leaking can complaints from consumers were, for all practical purposes, zero. Lost production from this plan amounted to an average of one line shut down every second day for 30 minutes. This calculates to 120 cans per minute × 30 minutes = 3,600 cans of lost production every other day, or 3,600 cans × 250/2 days = 450,000 cans lost production per year.

When the production manager studied this apparently expensive quality safeguard, he decided to eliminate the tester operator, and to bypass the storage operation. The advantages, according to the manager, amounted to $35,000 savings by eliminating the tester operator, increased production of 450,000 cans per year, and elimination of leaker downtime costs. A detailed analysis was then made as shown in Table 4.1.

Although there were other possible savings connected with the plan to eliminate defect removal (such as mechanical labor to maintain and repair the storage unit, special handling and storage of defective production, training personnel in detecting causes of defects, etc.), the manager considered that estimating these costs might be subject to argument, thus weakening his case for this project.

Unfortunately, the manager had focused his attention on savings and

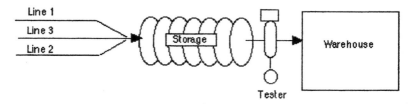

**FIGURE 4.2** Can line configuration.

*Table 4.1.  Cost of Leakage Detection and Removal ($).*

| | |
|---|---:|
| Scrap and salvage costs | |
| Cost of scrapped cans | |
| 64.8 leakers/hour × 2,000 hours × $0.12 /can | $15,552.00 |
| Scrapped cases | |
| (64.8 × 2,000)/18 cans/case × $0.48 per case | 3,456.00 |
| Salvage labor cost | |
| 2 laborers × $12.0305/hour × 400 hours | 9,624.40 |
| | |
| Downtime costs (nonproductive costs) | |
| Labor charges include accident, health, vacation, | |
| retirement funds, and other "perks" | |
| Labor and overhead costs per hour | |
| Line operator | $16.0800 |
| Mechanic 1/8 × 20.50 = | 2.5625 |
| 8 Process operators 1/3 × 17.25 = | 46.0000 |
| Can feeder 1/3 × 16.38 = | 5.4600 |
| Overhead per line | 8.0358 |
| Total, per hour | $78.1384 |
| Annual downtime cost (1/2 hour every other day) | |
| 250 days/year × 1/2 hour/day × $78.1384 = | 9,767.30 |
| Saving of tester operator wages | 35,000.00 |
| Total estimated yearly savings | $73,399.70 |

had overlooked the customers' expectation of purchasing nonleaking cans. Without the storage and the tester, it might be impossible to detect and remove leakers from the production stream. In time, an undetected cause for defects might generate leakers at a rate far exceeding the 0.3% currently produced. Worse still, under the most favorable of conditions, this process generates up to 64.8 leakers per hour for 2,000 hours each year—a total of 129,600 unhappy customers.

Not included in the manager's calculations was the cost to replace 129,600 customers each year. Unbelievable though it might seem, the plan was accepted, placing an enormous burden on the marketing department. Will this company develop another strategy to remain in business? Only time will tell.

What is the most effective next move for the quality manager? Under these conditions, probably doing nothing for a little while is the best idea. The quality decision has obviously been made over his head, and an "I told you so" attitude would be received with hostility. By quietly working with the engineering department (if one exists), it might be possible to study the possibility of introducing an alternate leakage detector system, such as a sonic device. It would be well to allow credit for this new idea to remain with the engineering department, since an embarrassed upper management might not look too kindly at the quality manager's "caving in" to the idea of allowing leakers to be shipped, whether true or not. An engineering solution to the production manager's seriously flawed economic "improvement" appears to be the most satisfactory approach to take.

A stopgap system to reduce the possibility of shipping leaking cans, although far less effective than the existing system, might be installation of a batch system of isolating production into identifiable lots. These could be tested after the fact to determine if they were satisfactory to ship. Of course, it might be costly to operate such a system. Production would set aside x cases each half hour for quality control inspection; pallet loads would be identified by the time produced, using pallet tickets; Quality Control would provide personnel to inspect the set-aside samples, perhaps every half hour; warehouse personnel might be required to extricate pallet loads found by the quality control tests to be suspect; labor would be delegated to break down the pallet load, open the cases, allowing each can to be checked for leaks; and, finally, the cases of sound cans would have to be repacked.

Under this plan, if a lot were found defective, all failed cans (and this could be a sizable number) would have to be salvaged, cans destroyed, and the product blended back into the production line, if possible. The do-nothing alternative? Don't perform any inspection, and wait for the customers to find the leakers. The cost of this alternative should be obvious.

## An Automatic Blending System

A small company developed a new product requiring weighing and blending 23 separate ingredients. It was consumer tested and found to be a potential winner. A temporary mixing system was devised, using hoppers, a floor scale, and wheeled bins that were fed to a batch mixer (see Figure 4.3).

After a year of highly successful test marketing, the company owner had an outside engineering firm design an automatic blending system in which 23 metering screws fed a screw conveyor that was expected to blend the materials as they were delivered to the fillers. As one

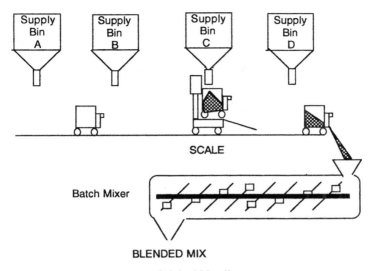

**FIGURE 4.3**  Original blending system.

might expect after looking at the system (see Figure 4.4), the conveyor had no opportunity to blend the material from the last few metering screws, and the very first production run using the new equipment was rejected immediately by Quality Control. (In the interest of clarity, only four of the 23 bins and metering screws are shown.)

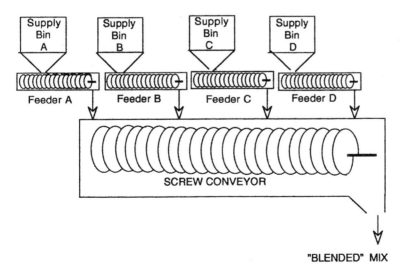

**FIGURE 4.4**  "Improved" system.

Obviously, this problem could have been avoided if the owner had thought to allow others in the company to evaluate the engineering drawings before installation. In fact, if the owner's thoughts had been presented on the back of an envelope to the production or quality people before the engineering concern took over, someone would have been sure to have spotted the design error. For something this obvious, even the shipping clerk could have suspected it would not have worked.

Realizing that future production system design should be discussed with his people, the owner belatedly called on the plant engineer to ask for suggestions that might solve the dilemma. As luck would have it, the old batch blender was still in the company warehouse, and, during the meeting, the plant engineer suggested reinstalling it at the end of the new conveyor (see Figure 4.5). After a few minor adjustments, mostly involving conveyor speeds, the corrected improved system worked satisfactorily.

There was no call for a statistical analysis of this mixing problem.

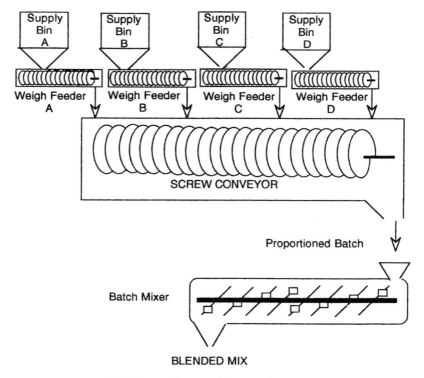

**FIGURE 4.5** Corrected improved system.

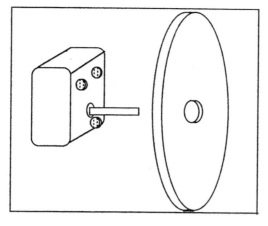

**FIGURE 4.6.**

The difficulty was in the process design, and the solution was found through reasoning and intuition based on experience.

## The Intermittent Defect

The parts for an electric clock motor were manufactured by various suppliers and assembled in a small plant by five long-time employees. The motors were then cemented onto polished agate rock slabs, with the hour hand and minute hand drive shafts protruding through a predrilled hole. Once in a while, the completed clock refused to run because of some kind of jam in the drive shafts (Figure 4.6).

The shop owner first suspected careless workmanship and set up a simple system by which each assembly was stamped with a worker's identifying number. This proved to be fruitless. On some days no defects were created. On some days a few defects were observed, and all employees' identifications were found scattered through the day's production in approximately equal numbers. Occasionally, the entire day's lot of motors was defective, and again, there was no indication that any one worker was responsible.

The owner then considered several different approaches: perhaps the epoxy cement used to fasten the motors to the slabs was dripping onto the shafts; possibly there was dust blowing through the assembly area on hot days when the windows were open; maybe the holes drilled through the rock produced burrs that jammed the shafts; or the reason might have been damage caused while assembling the hands to the shafts.

Purely by accident, the focus shifted to parts suppliers. The first clue to the mystery resulted when the shipment of Hatchi clock cases was held up at customs, and the only parts available were the Acme supply cases. For two days, until all of the Acme cases were used up, the defect rate was disastrous. When the Hatchi cases arrived on the following day, they were used exclusively, and, miraculously, no defects were found. With this information at hand, it was no more than a matter of minutes before the owner found that the Acme drive shaft hole was stamped slightly off center and that the shafts were binding on the case when installed.

This may seem to be a trivial problem (except to the shop owner), and, knowing the solution, several analytical techniques come to mind that might have solved the problem more quickly. On the other hand, it is doubtful that a statistical analysis would have provided a faster solution until the key problem was recognized; and at that point, no mathematical analysis would be necessary.

### A "New and Improved Cleaning Powder"

For over eight years, one of the major moneymakers for a chemical supply house was a one-pound box of trisodium phosphate powder, marketed as the finest frypan cleanser available. The package logo was modified occasionally, but the product and the package configuration remained unchanged. After many trials, the research department developed an improved powder that contained a detergent additive, a colorant, and a mild fragrance. The marketing department, working in secret so that competition would not know of the plans, redesigned the label but retained the same package configuration. They also secretly arranged to order the first-run supplies. Without prior notice, Marketing called a meeting of all of the company's department heads, at which time they unveiled the new package and announced that production would commence in one week.

The startled processing manager, the packaging line supervisor, and the quality control manager peppered the marketing people with a barrage of questions: "How do we clean the dye out of the equipment?" "Do we know how long to blend this stuff?" "Will it fit in the box?" "Is the dye FDA approved?" "There's probably going to be a dust problem, and we'll have to run slower. Is that OK?" "Are there any safety hazards we should know about?" "Will the mixture segregate itself on the production line?". It soon became apparent that using the high-handed secretive approach by the marketing group was the wrong way to improve the product line.

Some of these questions were immediately answered by the research people, but some would have to await the first production run. As it turned out, things ran rather smoothly right up to the packaging line, where it was observed that the new product had a marginally lighter bulk density, and one pound overfilled the box.

As in the previous examples, statistics had no part in curing this ill, but suggestions poured in from the quality engineers. The two top suggestions were selected: slow the line with added vibration and order larger boxes. The first of these suggestions permitted filling initial trade orders, and the second subsequently solved the problem.

### The Jammed Powder Conveyor

A new powder conveyor was installed without considering the carryback of dust on the return side of the belt. As a result, excessive powder buildup on the lower housing (see Figure 4.7) caused the belt to drag.

During the very first day's operation, the resulting friction produced extreme heat and eventually smoke. Fortunately, the motor drive overloaded before a fire started, and the conveyor shut itself down. The solution is shown in Figure 4.8 where the bottom of the housing was enlarged to create a self-cleaning hopper, allowing removal of excess dust.

Unlike the cleaning powder project discussed above, the design for this conveyor had been carefully analyzed by both production and quality personnel. Much of the design considerations were concerned with the possibility of contamination from bearing lubricants, as well as the probability of the bearings becoming inoperable from product

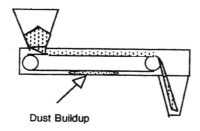

Dust Buildup

FIGURE 4.7 Jammed powder conveyor.

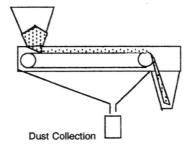

**FIGURE 4.8**  The modified powder conveyor.

contact. Consequently, attention to the possibility of carryback on the lower side of the belt was overlooked. At a subsequent meeting, the simple solution to the dust problem was devised.

Note how the last two process problems were uncovered and solved by the use of experience and intuition—without reference to statistics. It should again be emphasized that these examples in no way show limitations to the statistical approach to quality control and process improvement, but that the Deming philosophy of always reducing every problem to numbers may not be universally required.

### Net Weight Control

Usually, weight control projects require statistical analysis, and some very powerful techniques are available to produce enormous dollar savings in reduction of overpack. The following example is somewhat of an exception, since the process we shall examine was out of control, and the end-product weight had historically been regulated by guesswork.

One hundred consecutive samples of the one-pound containers were removed from the production line and weighed to the nearest 0.1 ounce. A cursory examination of the data confirmed that guesswork was a useless process control tool. A simple graph was prepared, and, as shown in Figure 4.9, all of the weights were above the label weight of 16.0 ounces.

The weight variations appeared to be uncontrollable, with periodic container weights excessively high. An interesting side issue is the observation that more than 1.5 ounces in excess of label claim were occurring regularly. This would suggest that the empty container must have been considerably oversized and presumably cost more than necessary.

This wasteful process probably could have been studied and improved through the use of statistically designed experiments and the application of X-bar and R charts. However, the process was so badly out of control that immediate remedial steps were indicated.

Before even considering the possibility of instituting statistical weight control on this line, it was necessary to examine the production line from the feed hoppers, scales, and open container conveyor system to locate obvious causes of the erratic performance.

As indicated by the cyclical high weight on the graph, one of the scales was considerably out of adjustment and was delivering approximately 17.25 ounces. On further study, it was found that none of the scale bearings had been sharpened or replaced in over three years, and the result was erratic performance. Four of the counterweight adjusting screws had been lost and had been replaced by poorly fitting combinations of wires, nuts, and washers, further adding to the irregular operation. One of the scale hoppers was out of line, and would fail to dump until the operator tapped it with a metal rod. A sharp turn on the filled container conveyor produced severe vibration, causing some of the product to be thrown from one container to the next.

When all of these engineering problems were corrected, another set of 100 consecutive samples was withdrawn and weighed as above. The resulting graph (Figure 4.10) showed a remarkable improvement.

The range of weights has shrunk from about 1.5 ounces to about 0.4 ounces, but note that the average net weight has not changed from the previous chart. The label weight (shown on the chart as "Target") is 16.0 ounces, and, in both of the above cases, the average output from this line is about 0.75 ounces too high. The costs associated with this overpack certainly affect the balance sheet adversely.

The next step must appear to be obvious: lower the target weight by

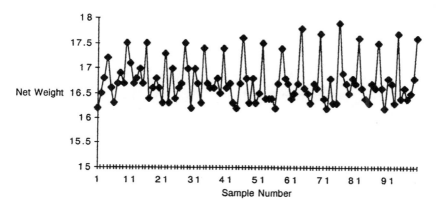

**FIGURE 4.9.**

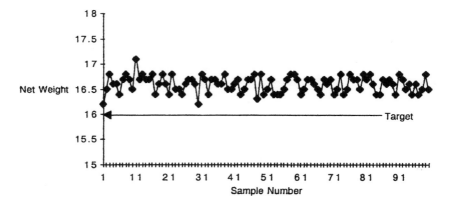

**FIGURE 4.10.**

0.75 ounces. When this is done, the performance is vastly improved, as shown in Figure 4.11.

Up to this point, simple observation, common sense, ingenuity, good housekeeping practices, and a bit of engineering have eliminated the major sources of variation, but all of these efforts will be lost unless control techniques are installed immediately. *Now is the time to institute statistical quality control!*

By referring to the government standards available in such manuals as Handbook 133, and applying some relatively simple statistics, it is possible to set the target weight at a level that will satisfy the management's arbitrary lower limits and reliability requirements. This can be accomplished with a known probability of success that will ensure that

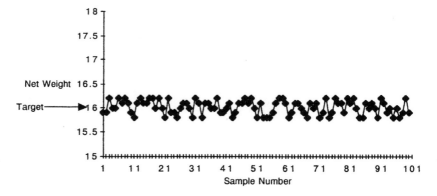

**FIGURE 4.11.**

governmental weight requirements are being met at least cost to the company.

However—yes, there are some limitations. The weight system that was studied in detail above must be monitored on a routine basis. A strict maintenance program must be instituted and adhered to rigorously. A sampling system that defines the frequency of sampling, the number of samples, the manner in which they are to be selected, the exact technique for weighing, and how they are to be charted, reported, and acted upon must be determined by statistical analysis. This will not only ensure weight control but will also point the way to even further improvement by proper interpretation of the control charts.

## *Eliminating Process Fires in a Polymer Process*

At the final process step, vats of a granular product were heated in an oil bath until polymerization was complete. At the precise moment of complete reaction, the product was then propelled by air pressure into a cooling centrifuge, which stabilized the product and cooled it for the mixing step which followed. The final moment of the cooking reaction was controlled visually by the operator, using a metal color plate that showed the precise brown shade of the completely polymerized mix (Figure 4.12).

Unfortunately, control of the heating step was critical, and, when not carefully watched by the operator, the entire batch could catch fire.

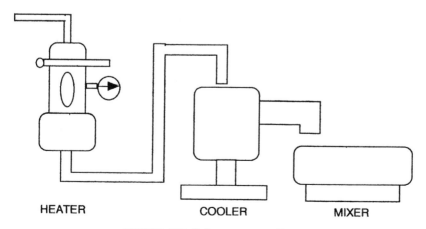

**HEATER**                    **COOLER**              **MIXER**

**FIGURE 4.12** Polymer process flow.

Loss of the materials was serious enough, but compounding this was the loss of production time because of equipment cleanup and evacuation of the production area until the fumes could be cleared out of the building. On some occasions fire detection equipment was activated, bringing in the local fire department and adding further delays in placing the line back in operation. The frequency of these disastrous fires was about one every three or four months. When the process was first transferred from the pilot plant, this danger was well known, and, as a precautionary step, the oil bath was equipped with a temperature-sensing and display device that contained an automatic shutoff mechanism in case the temperature approached the flash point. Unfortunately, by the time this temperature was reached, it would be too late. The process is exothermic in its final stages, and the flash point would be quickly reached.

The engineer assigned to correct this serious problem considered dozens of possible solutions before he decided that, like so many process difficulties he had experienced in the past, manual control of a critical operation is unreliable, and should be replaced with automation if possible. The final control step had been a manual one: the skilled operator compared the color of the batch to a hand-held sample of the target color of a finished lot. The engineer researched available automatic color controllers and selected one that continuously scanned the batch as it was being heated, displaying temperature readings on a large dial, sounding a continuous buzzer as the critical temperature was being approached, and automatically cutting off the heat at the precise moment of the completed reaction.

Over the next two years, there were no more fires from overheating. As a bonus, there were no more underprocessed batches that would have had to be reworked by blending.

## Tea Bag Quality Improvement

The improvements in the method of making a cup of tea were originally customer driven, but some of the later changes were cost or marketing motivated in an attempt to present some distinguishing feature to each manufacturer's product. During the 1950s, the industry's quality control efforts received a severe jolt that forever changed supplier/customer relationships.

But, first, a brief review of the history must be given. Tea leaves (1)

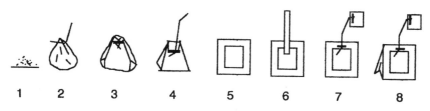

**FIGURE 4.13** Teabag design quality improvement.

were originally brewed in hot water directly in a cup or a pot. To avoid the unpleasant mess of wet tea leaves, metal or ceramic strainers of various sizes and shapes were invented. Some of the strainers were cleverly designed as clamshell shaped or screw-topped. The first tea bags (2) were made of lightweight, loosely woven cloth mesh and were hand-filled and hand-tied with light cotton string. The consumers may have considered this novelty an improvement, but wet cloth did nothing to improve the flavor of the tea (Figure 4.13).

Eventually, a filter paper was developed that was thin enough to permit fairly rapid diffusion of the brew, and was flexible enough to be formed into a pouch (3), hand filled, and stapled closed. The shape of the bag changed as various machines were invented to replace the hand operations. Machine design also permitted stapling a string to the bag (4), thus answering the customer's need for an easy method of removing the wet bag from the cup. It took several years before a lightweight, porous, heat-sealing paper was developed to the point where it could be tried on the public. There was no consumer demand for this modification (5), but it provided a means of high-speed, lower labor cost production. Various modifications have appeared since: heat-sealed plastic tags (6), stapled string with tag (7), and pouch-type bags of several designs (8).

A major quality disaster struck the industry without warning in the 1950s, when a flood of customer complaints overwhelmed major tea bag manufacturers. Within a single week, thousands of letters described how the tea bags fell apart in the cup or, worse still, as they were being removed from the cup. The cause was found to be a reduction in the amount of wet-strength binder in the paper. The reason for the change was that the major paper supplier unilaterally decided to lower his production cost by reducing the percentage of this expensive chemical binder—without notification to the tea bag manufacturers. Recalling several weeks of production in the retail pipeline proved to be a seven-figure disaster.

Several new quality control principles were quickly established as a result of this experience:

1. Never assume that a product is satisfactory when it passes all of the normal production line quality checks. *Always* test a sample the way the customer will use the product.
2. Maintain a close relationship with the suppliers. This experience ended the unqualified trust in the supplier's integrity and opened the door to supplier/purchaser partnering.
3. Just to be sure—always test the raw material shipments as they arrive from the suppliers. If possible, try some on the production line.

### Preventing Miscounts

The new food service packaging line for a one-pound soup mix pouch was a disappointment from the day it was installed. Product would fall into the heat seal, causing leakers. The film material used would stretch and tear. Weight control was intermittent, and poor at best. Electrical failures in the complex timing devices of the machine occurred daily. Employees were talking about the need for higher wages because the work was difficult and the output was much higher than the prior hand-filling operation (when the machine was running). Nearly every hour the film would drift out of alignment, resulting in pouch leakage (Figure 4.14).

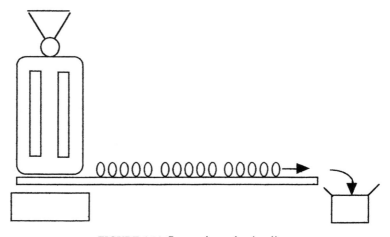

**FIGURE 4.14** Soup mix packaging line.

Over several months, with the help of the machine manufacturer, the maintenance shop, and the line supervisor, all of these problems were cured, one by one. Now that the line was running smoothly, the focus shifted to consumer acceptance of the new pouch design. The quality control manager, who previously had been besieged with customer complaints about leaking bags and underweights, reported to the management that the only remaining serious problem was short case counts.

The product flow following the pouch-forming/filling machine was a simple one, consisting of a 15-foot conveyor belt, at the end of which a worker counted the required number of pouches (25), dropped them into an empty case, sealed it, and stacked it in a pile on the floor. Although this seems simple enough, the case packer was required to set up the empty case, seal the bottom using a tape dispenser, remove and stack filled cases, and concentrate on the proper count as the pouches streamed down the line. As a quick fix, the maintenance department fitted the discharge conveyor belt with a Geneva drive, which momentarily speeded up the belt every fifth pouch. This provided a space between groups of five pouches each so that the case packer at the end of the line would have a simpler job of counting the number of pouches to fill the case. It would have been the perfect solution if only the line ran continuously, if the pouches would not occasionally fall off the line, and if the packer were not distracted by his other duties.

The quality control manager watched the process for a few hours and came up with a simple solution: place a platform scale equipped with a tare mechanism and a dial readout at the end of the conveyor belt. This would allow the case to be checked for the correct weight and, presumably, the correct count before sealing. To demonstrate, he borrowed a scale from a manufacturer, set the tare to the weight of the empty case and 25 empty pouches, and placed an empty case at the end of the line on the scale so that the pouches would drop into it. As soon as the dial read 25 pounds, the case was removed and sealed.

As happens so often in the real world, suggesting a solution, demonstrating the solution, and installing the solution is not always a simple process. Personality conflicts entered into the picture. According to the line supervisor, "Quality Control is supposed to run tests and write reports, not mess around with my line. And anyhow, who's going to pay for a $400 scale?" The machine shop was not too pleased to see what looked like a more successful answer to the problem than their Geneva gear idea.

Each company has its own methods of introducing new procedures. In this case, the quality manager talked over the idea of the tare scale with the soup mix product sales manager at a friendly lunch. Within a few days, the sales manager suggested to the production manager that perhaps a tare scale at the end of the line would eliminate miscounts. That's all it took to eliminate the problem! Who got credit for the idea? Does it really matter? Chances are that the entire company knew where the idea originated.

### The Declassifying Filler

Over the years, customers would frequently complain about excessive sugar in their packages of hot cocoa mix. In fact, one irate consumer returned a package that appeared to contain nearly 100% sugar. With the hope of finding a pattern in the process that might explain this defect, consecutive samples of finished product were removed from the production line while operating at normal speed and examined for excessive sugar crystals. The samples were numbered and identified by filler head. The examination was accomplished by a simple test consisting of placing a weighed well-mixed sample in a vibrating screen device for a fixed time and weighing the sugar sifted through screens of the appropriate size mesh (Figure 4.15).

It became apparent that all of the samples from scales 1 and 6 contained excessive amounts of sugar. The results of the test were shown to the foreman and the line operator. The line operator almost immediately said, "I knew this. Every day when I ran my half-hourly check-weigh samples, the end fillers always looked whiter."

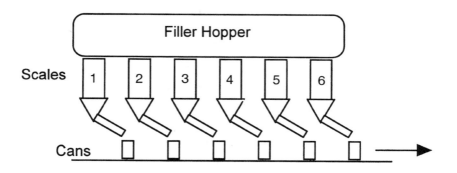

**FIGURE 4.15.**

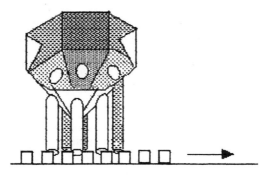

**FIGURE 4.16** Cutaway of pyramidal hopper.

Foreman: "Why didn't you say something about it?"
Operator: "Nobody ever asked."

The problem of "unmixing," sometimes referred to as "product classifying" or "segregation," is well known to process engineers. Depending on the product, either coarse particles or fines tend to work their way to the outside of hoppers during either flow or hopper vibration. To solve the problem, a hexagonal hopper was designed and constructed in such a way that all six sides received an equal amount of classified material as it flowed to the packages below (see Figure 4.16). Subsequent tests showed no further problem with segregation.

Although this example was selected to illustrate another case where intuition and experience were sufficient to solve a process problem, there are two other lessons to be learned. The fact that this problem had persisted for years is inexcusable. It should have been brought to the attention of one of the company's decision makers when first detected. Lesson 1: whoever was responsible for reading consumer complaints should have been instructed to pass such information along immediately. Lesson 2: whoever was checkweighing on the production line should have been instructed that reporting potential quality problems should be considered part of the job.

At the beginning of this chapter, it was suggested that many problems can be solved without the use of statistical quality control calculations, and statistically designed experiments are not always called for. In none of the examples selected above would the use of high-powered statistics have helped. In the case of the declassifying filler, a carefully designed factorial testing procedure that explored line speed, package size, product temperature, blending procedures, and other potential causes would have been costly, time consuming, and perhaps even fruitless in the event that the serial nature of the problem was not considered one of the potential factors.

As we shall see in the next chapter, there are yet other methods of analyzing process and product improvement and control that either ignore or make only tangential use of statistics. These other methods have received wide acceptance because they are readily understood and applied and because they are frequently successful for some classes of problems. This is not intended to belittle the proper use of statistics in quality control and process improvement. Some problems are unsolvable without its use, and the successful hour-to-hour quality control systems that fail to make use of statistics as the underlying principle are unlikely to succeed. Unfortunately, there are no set rules governing which situations can be solved intuitively and which require statistics.

# CHAPTER 5

# Quality and Process Solutions: Nonstatistical Methods

## INTRODUCTION

QUALITY problems can erupt at any point along the business process: in research and development, purchasing, raw materials receiving and storage, processing, finished goods storage, shipping, outside warehousing, marketing and sales, customer usage, maintenance, accounting, service, engineering, personnel—anywhere. It is unlikely, even in a small company, that a quality control manager can be found with the training, experience, drive, and time to successfully solve quality problems in all of these areas. Recognizing the need to develop a universal technique for quality problem solving, companies have tried various methods. Some companies have looked to the outside for help. They may call in an equipment manufacturer to suggest solutions or enlist the assistance of consultants to create new quality systems.

"Solution to quality problems" is an all-encompassing descriptor that is intended to include product or service quality improvement, process improvement, work simplification, scrap (waste) and rework reduction, error elimination, and discovery of new opportunities for process, product, marketing, or consumer satisfaction. Although all of the above appear to be somewhat quality related, the ultimate goal for solution to these problems is cost reduction (or profit enhancement). This comes close to the job description for *industrial engineering*: development and application of efficient cost and work standards for the various operations involved in processes.

An increasingly common procedure is to look to talent within the company. Where special abilities exist, a "troubleshooter" might be selected from the engineering department (if there is one), or from the area where the problem arises, to study and solve a specific problem. There has been some success with empowering each employee with the responsibility for solving his own quality problems. More often, groups of employees have been formed using a variety of techniques:

1. Establish a permanent problem-solving group with the authority to work in any area of the business.
2. Create teams for problem solving within individual departments.
3. Assemble committees to discuss specific problems as they arise.
4. Provide motivation through the use of descriptive names for the problem-solving groups: quality circles, transformation modules, quality teams, quality improvement committees, etc.

Any one of the group efforts listed above can be successful if there is management support and if the members are carefully selected and trained. Although statistical knowledge would greatly improve the chances for success of any of these groups, it is not absolutely necessary. In the event that statistical analysis is required, chances are that it can be performed for the team by others within the company. Many problems, if carefully selected techniques are applied, can be solved by reason and logic, without the use of statistics. Several of these will be discussed in some detail later in this chapter. For lack of a better term, we shall use the word "team" when referring to the circles, groups, or committees.

The concept of quality "circles" probably originated in Japan and has been reported by Japanese companies to be a successful technique for solving many types of their quality and process problems. However, when applied to American industries, the number of failures of quality circle attempts outnumbered the successes to such an extent that they gradually disappeared from the scene. The "circle" idea was reintroduced under the name "quality teams," and many publicized results were remarkably successful.

The choice of the word "team" may have been unfortunate, because it has so many conflicting meanings. "Two or more animals harnessed together to pull a wagon or plow" implies a single driver who directs the team's every move. "Two or more players engaged in a sport in competition with another team" bring to mind a football team where each player on the team has a specific function and is directed (without discussion) by a coach. "A team player" has been used over many years in the business world to imply that an employee performed his tasks exactly as management instructed him to do it. "A relay team" is where each runner completes one lap and hands over the baton to the next team member to complete the task. None of these definitions can be successfully applied to quality teams. Quality teams must *share* concepts and responsibility for their collective task. They may or may not have a leader, a spokesperson, or facilitator, and they must be empowered to implement their findings and decisions. The team members may all come from the same department, or they may each have specialized positions for specific functions in the organization. One would expect that there are no fixed rules for team composition; on the contrary, there are dozens of precise groups of rules formulated by various seminar instructors and educational organizations across the country. We shall examine a few of these principles.

## TEAM SELECTION AND TRAINING

Perhaps many of the improvement examples shown in Chapter 4 could have been discovered and created by teams, circles, committees, etc., but there is an extreme danger of "too little knowledge" if these teams are not sufficiently trained.

Team training, at the very least, should include explanations of the principles of team member selection, nonstatistical techniques in problem solving (brainstorming, process diagrams, fishbone analysis, Pareto diagrams, scatter diagrams, plan-do-check-act cycle, rules for interpersonal relationships), and some of the bare basics of control chart techniques. Teams should be able to understand the power of control charts, even though they may have to depend upon nonmember company personnel to construct the charts for them. The primary risk associated with an untrained team is that it tends to rush into a project with enthusiasm, choose a solution in haste, and stumble badly. This has been referred to as the "ready, fire, aim" syndrome. An untrained team will tend to lean heavily on the suggestions of a few vocal members, thus defeating the basic principle of team operation.

Some managers might believe that their processes and products can't be improved because their formal staffs have a "you can't get there from here" mind-set. This is the appropriate time to call in those most familiar with the process and the product: the hourly workers. Who should be on the team, and how should they be selected? There is obviously no one answer to this. In some highly complex processes, the workers will need occasional guidance from an experienced technical employee, and they should be included as members. If the project selected has a cost reduction goal, perhaps a representative from the accounting department should be available. Industrial relations is often required to discuss labor implications of projects under study. Many companies feel the necessity for a "facilitator" to be present at all meetings of the team. A poorly trained facilitator will stifle creativity and cooperation. Again, there is no universal decision regarding this need; each case must be studied on its own merits.

In some large companies, establishment of teams has been accomplished by a special committee (council or quality improvement group) that lays out the groundwork: need for a team, budget, size of teams, meeting places, composition, time tables, oversight, reward systems, and method of selection. This committee can act as a sounding board for team project progress reports and final implementation plans. Although this committee might provide some useful function in a large carefully structured company, it would be more apt to impede progress in smaller organizations.

Selection of team members can be by management appointment. This carries a certain air of approval and can prove to be an advantage to the team. By contrast, a team elected by line workers might wonder if the members were selected because of popularity rather than ability. In either case, a system of team member rotation can gradually change the complexion of its members so that it can eventually represent a cross section of the employees. Formation of a number of teams might assure management that some will be better able to solve problems than others, and over time, the number of teams can be honed to an effective few. The main difficulty with multiple teams is the creation of an environment of competition between teams. At first, this might be considered an advantage, but experience has shown that competition can generate jealousy and ill will. Under extreme circumstances, jealousy can lead to "faking the data" to find a solution before the other team does or "tripping up" the competition so that they will fail in their endeavor. Both of these possibilities are extremely destructive to successful team operation.

At two or three levels of a corporation, selection of teams is predetermined. The board of directors presumably acts as a team to explore business opportunities, strategies, finance, and direction. A major stockholder might be admitted to this circle, and operational officers might be summoned for their expertise on occasion. Otherwise, that team is unchanging. For the most part, the same situation exists at the next level of management: the vice presidents. It is conceivable that a vice president conference called by the president might include an invitation for a manager to appear for specialized input, but otherwise the team is a closed group. The structure of teams formed at the managerial level is far more flexible. It can be horizontal, consisting exclusively of managers, or vertical, created by members from a single department. Below the managerial level, teams may be homogeneous, including supervisors, workers, clerks, assistants, operators, etc. Of course, there are exceptions to this simplified classification of teams by position in a corporation, but they are not common. After all, a vice-president as a member of a team of line operators would cause uneasiness merely by his presence. Worse still, a vice-presidential suggestion to a team would come across as an edict rather than a starting point for discussion.

In small companies, where informality is more the rule, quality/process improvement teams have always existed on a casual basis with little regard for differences in job level. For example, "Hey Phil, here's a bunch of rotary trimmer blades for you that should last until the end of your shift. Pete and I were just talking about this, and we thought we could deliver the blades to you as we move the truck by your station. That way you won't have to shut down the line and get your own." When time permits, formalized teams of employees at any level can be chosen to solve specific problems. Large company or small, teams should consist of trained individuals who all cooperate to reach a common goal.

Special problems require a specialist on the team. If the problem involves engineering, pallet movement, or bacteriology, then the team needs an engineer or a forklift operator or a bacteriologist. Not everyone plays the piano; not everyone is capable of playing the piano; if the team problem involves a piano, someone on the team had better be a piano player.

Some situations require team conformance to a single leader's plan, not team discussion and argument. For example, there are some places where teams or individual empowerment may be a totally unacceptable method for improving processes or quality: a performing symphony orchestra, a 100-meter racing sailboat in competition, a basketball team on the court, or most military units.

Certain specialized industries may benefit from the use of team problem solving only at the headquarters level, not in field operations. For example, a fast-food corporation with a thousand identical retail outlets cannot permit modification of retail operations at the local level. In the first place, the customer expects to purchase the exact same product quality at any outlet, whether it be across the town or in another state. Secondly, if one of the outlets modified any of the carefully researched maintenance or cooking procedures, storage temperatures, or sanitation practices, based on ideas of a local team, the change might turn out to be illegal, dangerous, damaging, or costly. A third problem is that local changes to standardized headquarters-generated procedures are likely to result in loss of companywide product and service quality control, and possibly cost control as well. Ideas for improvement can be handled through field coordinators who are responsible for the operations of a number of outlets in a given region. This has been successfully accomplished by periodically assembling the regional outlet managers to perform as a product/process improvement team, with suggested improvements being forwarded to headquarters.

Teams have been classified by the type of goal they are expected to reach.

- Specific problem—everyone knows it's wrong, but nobody has had the time to fix it.
- General problem—sometimes costly interruptions or mistakes occur in the same process area, and it would be profitable to correct them.
- Product R&D—what other kind of service could the company perform that would be of value to the customer? How could the product be modified to add to our sales base? What new product could the company invent and produce?
- Quality R&D—how can controls be tightened so that there is less scrap, rework, customer complaints, downtime, mistakes, and associated costs?
- Physical process improvement—what process flow changes can improve output and reduce costs? What modified or new equipment would be an advantage?
- Administrative process or system improvement—study staffing requirements for operations. Investigate production scheduling efficiency. Improve flow of information between departments.

There is some debate over the need to charter a team in one of these categories rather than to give it free reign. Required to concentrate on a specific area, a team may produce faster results than if it were allowed to wander over the entire spectrum of opportunities. The obvious disadvantage to a restricted team is that it might accidentally stray from its course into a highly productive area and then be forced to abandon it.

## Team Training

Selecting a method for training is a decision best left to each individual company. Distributing training manuals without formalized classes rarely works—if ever. Teaching at off-site seminars or classrooms seems to be a method reserved for large companies with large training budgets. Classes given at the company by company trainers can be only as effective as the trainers used. A quality control manager or a quality control engineer who has received training at a team training seminar may be a successful instructor. Generally, one would expect a greater success record by employing professional trainers. They are available from quality control consultants, management consulting firms, or community college staff.

There are dozens of texts available on team training, and we will cover only the major concepts and tools: behavior, brainstorming, process flow charts, fishbone analysis (cause and effect), Pareto analysis, scatter diagrams, and the Deming Cycle. We shall look at each portion of the training program briefly, bearing in mind that volumes of literature are available for these subjects.

### Ground Rules for Behavior

The whole idea behind quality circles, process improvement teams, or committees is to promote interchange of ideas between people of different backgrounds, with varying capabilities, from departments with contrasting goals. Progress under these conditions can be fruitful only if the personal contacts are tolerant, respectful, and considerate—in short, well mannered.

We used to call it "manners"; later it was renamed "interpersonal relationships." It used to be taught at home and practiced by the family; for the past several years it has hardly been taught at all and is rarely considered during contacts with others. Worse still, lectures on "aggressiveness" and "assertiveness" seem to teach just the opposite: how to be abrasive. Some companies, in desperation, have resorted to special training programs supplied by seminars to reawaken their employees to the concept that they work with *people,* not just workstations, customers, suppliers, the maintenance department, quality control, and federal regulators, and that they need to learn to relate to these people with understanding, kindness, and good manners. Even in the academic world, the Massachusetts Institute of Technology recognized the need for rounding out the training of engineers by offering a course in manners, started in 1995.

Generalizing on human behavior is usually risky, since exceptions are easy to find. But, for the most part, this loss of the practice of manners training has been the result of major changes in the home and family environment. Under parental tutelage the children used to be taught how to eat, how to respect their parents, how to take turns with their brothers and sisters when they had an impulse to speak, how to smile at pleasantries, how to have compassion for others' difficulties, how to offer suggestions without offending, and how to disguise temper flareups—in short, how to get along with other people. There have been changes affecting the dinner table: some families have only one parent to guide dinner conversation. In many families, both parents work and come home too tired to face an hour of family training. Special television programs often shorten mealtime; in fact, television *at* mealtime virtually eliminates personal contact training.

The first few minutes of the initial meeting of teams must be used to outline the ground rules of good manners. Sarcastic, rude, or mean members should be counseled in private and removed from the group if manners do not improve. The ground rules might need repetition whenever infractions occur during the meetings. Without this continuous awareness by participants, it is unlikely that the group's effectiveness will surface, much less persist.

This is not meant to imply that team meetings should resemble a tea party. A successful team requires active (but not aggressive) discussion and continuous participation by all members. Both the reserved and the outspoken should be taught to be aware that they are partners and will share equally in the outcome of the team efforts.

### Brainstorming

The first activity in team operation is selection of a project. Even if the team has been presented with a specific project from management, it is often fruitful to start off with a clean slate and proceed through the brainstorming stage. The goal of project selection brainstorming is to solicit ideas from each member of the team for any project that comes to mind. Each idea is listed on a flip chart or blackboard by one of the team members selected as scribe.

"Scribe" is more effective than "leader," since a leader is likely to guide the thinking of the team, and an effective brainstorming session should be unstructured. In addition to the concept of a leader or a scribe, four other commonly used plans include the appointment of a facilitator, a spokesperson (presenter), a moderator, or a supervisor. The facilitator is a trained group leader who *facilitates*, rather than *contributes*, to the brainstorming sessions. The spokesperson acts as the leader but, in addition, acts as the sole contact between the team and management. A moderator keeps the activities focused. Appointment of a supervisor is thought to work well for some companies, since the supervisor is trained to lead and presumably has experience in guiding groups of diverse individuals toward a common goal. Perhaps the differences between the titles of the leader are of no consequence, but seminars on team operation seem to emphasize one or another as the most successful means of controlling teams.

No judgment of ideas is permitted at this stage—all ideas are recorded, be they deadly serious or frivolous, complex, unrelated, humorous, silly, or half-baked. Each team member should be given a turn to suggest a project; if no idea is immediately forthcoming, he should say "pass" while the remainder of the group contributes. The next time around, this member may have had a thought prompted by other offerings. Eventually, everyone should have had a few opportunities to contribute. The advantage of listing all ideas is that they remain in plain view on the flip chart for all to study, and the poorly thought-out suggestions will stand out glaringly as inconsequential. This tends to eliminate subsequent unlikely project suggestions.

Once the brainstorming process slows to a halt, the team can proceed to select the most promising project. There are many techniques for this. One of the easiest procedures is to label each suggestion with a rating from 1 to 5, with 1 being the most favorable. There is rarely a need to argue over these ratings, since team members have been instructed to look for neither credit nor blame. If there are several "1's," the team may then prioritize them, determining the order in which they could be worked on. The precise wording for the project selected should be recorded at the end of the project brainstorming procedure. This will eliminate the possibility of drifting during the analytical steps that follow.

The brainstorming process may also be used in construction of cause-and-effect diagrams and flowcharts discussed below.

### Process Flowchart

A process flowchart is a visual representation of all of the steps of a process, showing relationships between steps. It provides the team with a road map of the universe they are studying and may suggest location of problem areas, redundant procedures and loops, and areas of complexity that might be simplified.

The first time a team is asked to create a process flowchart, the idea is usually treated as an unnecessary waste of time. "Everyone on the team has been here for at least five years, and we all know how the process works." However, it is a rare process that every employee understands the same way. When one team member draws a flowchart of the process on the blackboard, somebody will suggest a change, an addition, or a correction, and then the flood of comments starts:

- "We don't always mix in polyethylene; sometimes we use urethane."
- "How about the unloading operation at the shipping dock?"
- "We've got three lines feeding the shredder, not two."
- "You forgot the step for punching in the computer."
- "We heat after the peeling, as well as before."
- "You left out the quality control check and approval at step 8."

Soon, everyone is involved. Eventually, the contributions stop, and the process is now understood by the entire team.

Flowcharts can be constructed formally, using industrial engineering symbols, simplified symbols, or squares and arrows. Figure 5.1 shows some of the symbols commonly used.

A simple example of a flowchart is shown in Figure 5.2. By the time a team is through working on additional details of this process, the diagram can become cluttered with many entries.

Additions to this process flowchart are probably required right at the incoming material step. For example,

- precontract sample testing
  Spindles
  Acetate solvent
  Linkages
  Discs
  Circuit board DCA

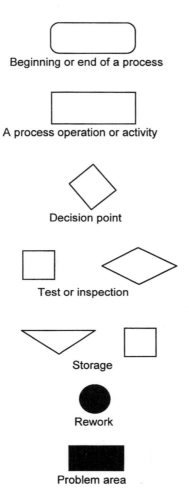

Beginning or end of a process

A process operation or activity

Decision point

Test or inspection

Storage

Rework

Problem area

**FIGURE 5.1** Process control chart symbols. (Note duplication by various authors.)

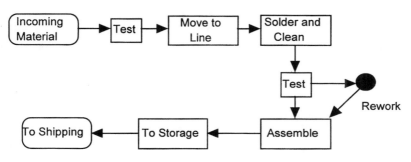

**FIGURE 5.2** Simple process flow chart.

       Circuit board amplifiers
       Aluminum chassis sheet
       Condensers and resistors
       Control cylinders
       Shippers
- Supplier quality control
       Plant audit
       Certification records
       Control charts

These and many other activities involving the suppliers would be shown at the left end of the flowchart, attached to "Incoming Material." The stages involved in handling, storage, and movement of incoming material should precede "Test." Such activities as stock rotation, damage prevention, segregation, and identification would then be attached to these three stages. In the end, the flowchart becomes extremely complex, but the team should have a better understanding of how the entire process works. Furthermore, hidden deep within one of these many side branches may lie the cause of the very problem the team is studying.

### Fishbone

The proper name for a fishbone diagram is cause-and-effect diagram. This differs from the process flowchart in several ways. It is highly structured at the outset and is subsequently detailed by the addition of experience, logic, and brainstorming on the part of the team. By contrast, brainstorming is not used in constructing the process control charts because these are intended to represent the process as it is currently practiced.

The fishbone diagram is built up from a series of causes, resulting in some specific effect. When first devised, the causes were assembled under four master headings: men, material, money, and methods. Over the years, these basic causes have been renamed and expanded. A "bare-bones" chart with the more common causes is shown in Figure 5.3.

In practice, a diagram such as that shown in Figure 5.3 is presented to a team after the project has been selected. If the project were "reduce the cost of Finnegan Pin production on line 17," the word "effect" is replaced by the problem statement, "line 17 production cost." A brainstorming session now proceeds, whereby the team ties all pos-

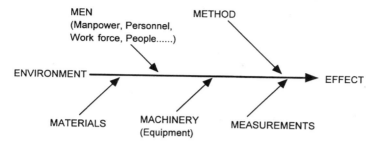

**FIGURE 5.3** Cause and effect diagram.

sible causes to the appropriate framework. Possible environmental causes, for example, might be time of day, lighting, temperature, atmospheric pressure, wind, dust, humidity, vibrations, noise, wall color, odors, etc. Brainstorming does not require justification of any of these suggestions; the unlikely ones are weeded out later. Sub-branches are introduced as needed. Vibration, for example, might be the result of the freight elevator, heavy floor machinery, lift truck traffic, bearing maintenance, machine imbalance, conveyor slippage, or other causes, and these might be attached to the vibration "bone."

Because the number of possible causes might be unwieldy, some teams are instructed to prepare checklists of causes between meetings. By compiling these lists before the next meeting, considerable time might be saved. As each cause is added to the diagram, the question "Why might this occur?" is asked of the team, and the responses are added to the diagram as further side branches. Untangling the intertwined branches and sub-branches can be challenging, and a number of procedures have been proposed in various articles and books on the subject. Some suggest tallying the causes in categories to locate the most common ones. This might provide a clue as to a solution. Another suggestion is to avoid overloading the diagram by isolating complex groups as they develop and preparing a separate diagram for those causes. To assist in wading through the huge volumes of entries, circling the more important causes will often help. Another suggestion to reduce the possibility of overload is to look for possible solutions to small problems as they appear in the network. This is likely to lift the team spirit as well as provide an opportunity for clearing out sections of the fishbone maze.

*Pareto Analysis*

A Pareto chart is a visual method of ranking data and is especially useful in prioritizing process areas for improvement. If defects are to be analyzed by causes, a Pareto chart of defects produced by a process, for example, will readily separate the "vital few" causes from the "trivial many." The technique is not limited to defect production; it has been used successfully to prioritize causes for machine downtime, clerical errors, injuries, customer complaints, unauthorized absences, school dropouts, vendor quality defects, inventory costs, size of customer orders, production delays, and innumerable other effects, both favorable and unfavorable.

To demonstrate how a Pareto chart is constructed, a simple example is shown in Table 5.1, showing how easy it is to prioritize defects found on an attaché case production line. The number of defects observed over a test period is listed in descending order with the most frequently observed defect at the top.

By plotting the number of defects versus type of defect (Figure 5.4), it becomes clear that the first five defects are the most numerous. But, when the cumulative percent of defects is plotted against the type of defect as shown in Figure 5.5, the first three defects are shown to represent nearly 80% of the defect population. In other words, if just these three defects could be eliminated, the total number of defects produced would be reduced by 80%. This shows the power of the Pareto analysis.

Figure 5.6 is another example showing the relationship of various defects in heating elements as they are removed from the final test station. By combining both the number of defects and the percentage data on a single graph, the relationship between the two becomes more clear.

*Table 5.1.*

| Defect Description | Number of Defects | Percent of Total Defects | Cumulative Percent |
|---|---|---|---|
| Scratch | 230 | 40 | 40 |
| Nick | 140 | 25 | 65 |
| Stain | 60 | 11 | 76 |
| Loose handle | 58 | 10 | 86 |
| Brocken lock | 45 | 8 | 94 |
| Sprung hinge | 10 | 2 | 96 |
| Missing key | 4 | 1 | 97 |
| Other | 24 | 3 | 100 |
| Total | 571 | 100 | — |

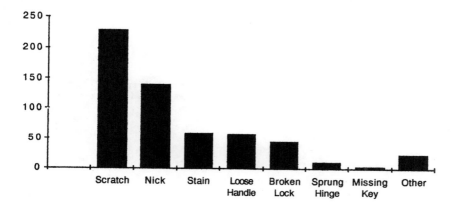

**FIGURE 5.4.**

Cumulative %

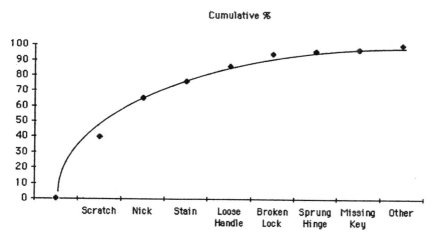

**FIGURE 5.5.**

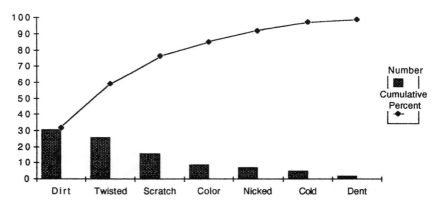

**FIGURE 5.6** Useful Pareto chart.

Logic would tell us that, once the causes for 80% of the defects have been removed, a subsequent Pareto chart would look much flatter, making it less useful (Figure 5.7). This might not occur on the second trial, since the next most relevant defect classes move up a notch; eventually, the number of classes of defects is so reduced that the chart becomes useless. Happily, reduction of the number of defect classes is the goal!

If the first Pareto chart of defect types is so flat that it requires 80% of the types to show 80% of the defects, a team is likely to abandon the concept of Pareto analysis (Figure 5.8).

Before giving up, the categorization (the x-axis) should be examined. Instead of "types of defects," perhaps another category, such as

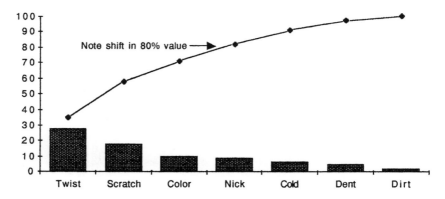

**FIGURE 5.7** Pareto chart showing dirt defects nearly eliminated.

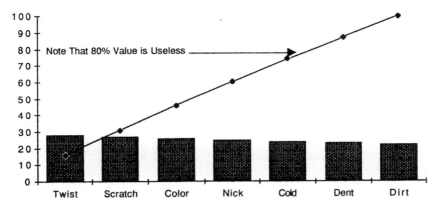

**FIGURE 5.8** Pareto chart where defects occur at nearly same frequency.

"operation at which defects occur," might provide the key to defect reduction. In the "useless Pareto chart" shown in Figure 5.8, if the individual defects were analyzed by machine or by line, it might show that a particular machine is at fault.

Before leaving the concept of the Pareto chart, it should be pointed out that there is more to problem solving than mere numbers. For example, even though "missing key" accounts for only 1% of the attaché case defects, it may be responsible for 98% of all of the consumer complaints received. Furthermore, it may also be the most important contributing factor to losing major department store accounts. In the food industry, contamination of the feed belt with *Salmonella* or *Clostridium botulinum* may appear at the far end of the x-axis of a Pareto chart at the 0.003% level but obviously is by far the most important defect to eliminate, irrespective of the 80% rule.

There are a few other important aspects to the use of Pareto charts. The scope of the data used for a Pareto chart should be considered carefully. It is possible to obtain a misleading picture by selecting data from a short run that might be out of control. For example, if data are collected from a short run in which raw material only from supplier A is used, defects (or costs, stoppages, etc.) might be significantly different than those from a run that includes materials from suppliers B, C, or D.

In the early stages of team operation, it might be desirable to solve one of the easy problems rather than one of those in the 80% category. Successful solution to the first problem area tackled by a team builds confidence. Chances are that, if the most frequently occurring problems had been easy to solve, they would have been eliminated long ago.

These factors must be considered by the team during the prioritizing phase of problem solving. Pareto analysis is an excellent aid to prioritizing, but it is not the last word!

## Scatter Diagram to Relate Two Variables

In common with the techniques described above, scatter diagram analysis of a problem is relatively simple to perform, requires no high-powered math, and is generally a pleasant experience. Conclusions are frequently almost immediate. A scatter diagram is a useful tool for investigating the relation between two variables selected at the same time. We will look at a simple example: the relation between the number of tellers at a bank and the number of customers served per hour. Data are collected over a period of three days. The numbers vary from day to day because of differences in the speed with which customers can be processed, the variations in bank traffic, and the time of sampling (Table 5.2).

A chart prepared from the above data shows a not-unexpected relationship: the more tellers, the more customers processed per hour (Figure 5.9). Even though the relationship is not precise, there is a definite trend shown, and an estimate can be made from the chart regarding the number of tellers required, above which no particular improvement in service is observed, in this case five tellers.

Note that there have been no calculations required to reach this tentative conclusion. The next steps are up to the team: *plan* an extended

*Table 5.2.*

| Number of Tellers | Number of Customers Served per Hour | | |
|---|---|---|---|
| | Monday | Tuesday | Wednesday |
| 1 | 15 | 14 | 12 |
| 2 | 18 | 16 | 17 |
| 3 | 24 | 10 | 18 |
| 4 | 6 | 22 | 23 |
| 5 | 30 | 29 | 28 |
| 6 | 33 | 31 | 30 |
| 7 | 27 | 32 | 30 |
| 8 | 31 | 29 | 34 |

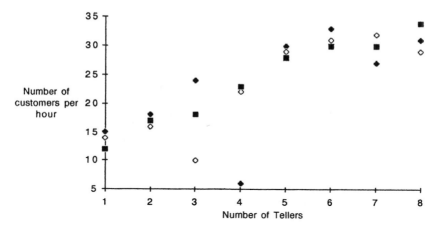

**FIGURE 5.9.**

test (bearing in mind the extra costs for improved customer service), *do* it, *check* the results, and *act* to either institute a permanent change in teller service or consider conducting further study, perhaps to improve the speed of teller activities.

Another simple example taken from a manufacturing operation study might be the examination of the relationship between line speed and defect production. It is entirely possible that a graph such as that Figure 5.10 could result, showing no apparent correlation.

How can this be? The line speed was increased from 1 per second to

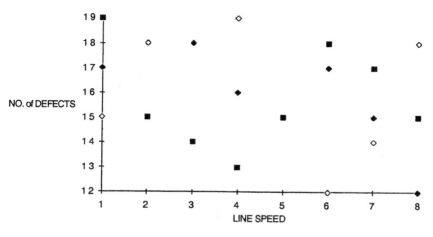

**FIGURE 5.10.**

8 per second, and the number of defects produced was unchanged; at least no pattern was observed. It could be some simple explanation such as the defects are caused by the operator, who removes one unit every 10 minutes, and, in the process of inspecting it, he damages the unit, thus causing a defect to occur at a constant rate, regardless of line speed. The reason for nonrelated variables might, in some instances, be of no importance but would merely serve to eliminate some possible improvement investigations from the list of candidates.

One final word on diagrams: there is another technique known as the sequential analysis chart. For this type of study, a series of measurements is made over time, with no intentional variation introduced into the process. On some occasions, a pattern of observations will repeat itself at a fairly consistent interval. This is another indication of some external factor operating on the system that should be identified and investigated. The effect can be a positive one (higher yield every Thursday at 3:15 P.M.) or a negative one (main circuit breaker blows whenever the freight elevator is operating).

## Plan/Do/Check/Act (The Deming Cycle)

The PDCA cycle is a formalized procedure for continuous improvement, sometimes referred to as the transformation model (Figure 5.11). It is a relatively simple concept that is intended to provide a structure for stimulating a team to produce a stream of uninterrupted quality and process improvements. Some of the major concepts of each step follow.

*Plan*—The first step of planning is project selection. Under some conditions, the project is preselected by management, but it more usually is the end result of a brainstorming session conducted by a team or a clear outcome from the ACT stage of the cycle. In the event that the team is confronted with a need to select from a group of possible projects, some procedure should be available for final evaluation. The usual criteria are cost savings, increased profits, and improvement of process, service, or quality. Secondary criteria might include availability of data, the cost to set up a test run, the need for floor space, or the likelihood of success. In the early stages of team operations, the time required for completion or the potential for success are vital criteria. There are also some nonmeasurable considerations to project selection, such as support of management and the union, specific mandatory

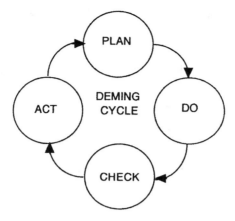

**FIGURE 5.11.**

criteria laid down by management, and probable reactions of employees.

The following principle has been found effective in avoiding endless debate during the project selection phase: aim for "the best solution," but accept "a good solution" with the understanding that continuous improvement will follow. This produces quicker results and avoids team discouragement.

After a plan is selected, clear definitions and procedures are required. The team must reach agreement on the goal. The steps to be taken during the project implementation should be diagrammed on a process flowchart; data sheets for process measurements should be prepared; procedures, timing, equipment, laboratory backup, and personnel required should be clearly defined.

Most plans will require some preliminary analysis of current procedures to establish benchmarks for comparison. For example, it might be desirable to study existing control charts and process capability data to assist in predicting goals.

*Do*—This is the project implementation step, and involves making some sort of a change in the standard procedures of a process as defined in the plan. The change should ideally be continued with sufficient data collection until a control chart indicates that the new process is under control. At this time it is usually desirable to construct a process capability chart as well. These will prove invaluable in analyzing the effectiveness of the change.

What should a team do when the implementation step indicates that perhaps we should have opened the valve a quarter turn more, or we should have asked for two more clerks at the checkout counter, or we should have used material from supplier X, or we should have shipped the products by air freight? As a general solution to this dilemma, the test run should be continued to completion exactly as planned, and, if conditions permit, the modification might be introduced at the end. Of greater importance, the team should recognize the inadequacy of the plan as it had been designed and use this information for future guidance. If conditions do not permit the modifications at the end of the test, they should be considered for a future improvement study.

*Check*—This step covers team analysis of the project data and determining if an improvement has been accomplished as originally planned. Before proceeding to the *act* step, however, final analysis of the results should be expanded to include effects on personnel, costs, environment, morale, safety, or any area not expressly covered in the original plan. Control charts generated by the project should be studied to determine if the control limits have changed and to identify outliers for further study. If the test failed to produce the improvement desired, the entire procedure should be reexamined to isolate the causes for the failure. Partial successes and unexpected problems should be documented and discussed for further examination or for use as a guide to future projects.

*Act*—If the project has indicated improved quality, process, or cost-effectiveness, it should now be implemented on a trial basis. Whatever process changes have been observed must now be standardized and documented. Line personnel (or others involved with the process change) must be adequately trained in the new procedures. New control charts should be instituted after a review of the sampling number, location, and frequency. This period of change should be treated as if it were a hospital patient who has been given a cure and is being monitored to assess improvement. A final report to management is advisable, with copies for the other teams.

*Plan*—The entire process of PDCA now starts over again.

## EXAMPLES OF TEAM SUCCESS

The use of team efforts for quality and process improvement has become increasingly popular over the past several years, and dozens of examples of successes appear in the literature. Not all efforts produce improvements, but these failures are rarely reported. A few examples of previously unpublished successes and failures follow.

## Materials Handling Study

First is an example of a team effort that proved highly successful. In a 30-year-old multistory plant, the entire process had been built around the principles of gravity flow and cheap labor. Although there had been no change in the law of gravity, cheap labor had become a thing of the past. On instructions from the operations manager, a team was formed to reduce labor requirements by finding ways of improving the process one step at a time. The first improvement was to be in freight car loading.

The team selected by the operations manager consisted of

1. The manager's assistant, a 10-year employee who had come up through the ranks, and was thoroughly familiar with the plant
2. An industrial engineer with 6 years' experience in engineering projects
3. The shop steward, a line supervisor and long-term employee
4. A warehouseman who had worked on all jobs where finished cases were stacked, conveyed, and car loaded

During the brainstorming sessions, a sketch of the material flow from the process to the freight car was drawn (see Figure 5.12).

This generated a number of questions regarding specific case movements that were not clearly shown on the sketch. To provide the detail needed, a flow diagram was constructed (see Figure 5.13).

It was suggested that the manual gravity conveyor system for bringing finished cases into the freight cars individually on gravity roller conveyors might be replaced by delivering forklift truckloads of pallet-

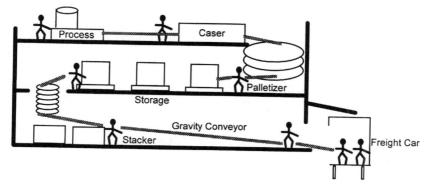

**FIGURE 5.12.**

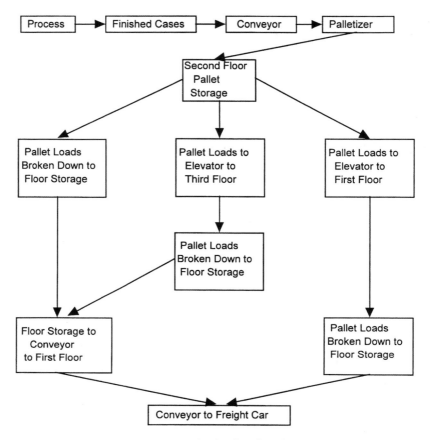

**FIGURE 5.13** Car loading flowchart.

ized cases directly into the freight cars. This could eliminate four of the six car loading warehousemen currently required. However, the old gravity system had dictated that the railroad tracks be installed two feet below the shipping floor level, and, if forklifts attempted to drive a load down an inclined platform into the freight cars, the pallet load was sure to tumble off the forks (see Figure 5.14).

More brainstorming suggested raising the track level to that of the shipping dock. The cost of this possibility proved prohibitive. A later session came up with the idea of having the forklift *back* into the freight car, thus eliminating the danger of losing the load (see Figure 5.15). This idea in turn led to the suggestion that clamp trucks be used, rather than forklifts, thus eliminating the need for pallets, and permitting machine loading of the freight cars (except for the doorway sec-

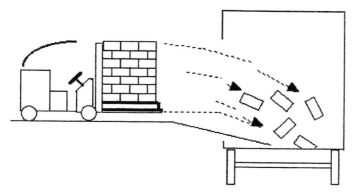

**FIGURE 5.14** Fork truck loading.

tion, see Figure 5.16). Clamp trucks are essentially the same as forklift trucks, except that the forks are replaced with opposing vertical clamp plates that are capable of squeezing and transporting stacks of cases up to about six feet high.

It wasn't long before the team looked into material handling further back into the plant and found that an entire handling system could be devised using clamp truck loading throughout. By developing a revised flowchart of the process (see Figure 5.17), the team determined the location and nature of new equipment required and was able to present a complete study to the operations manager.

The net result of these changes was the elimination of 22 of the 30

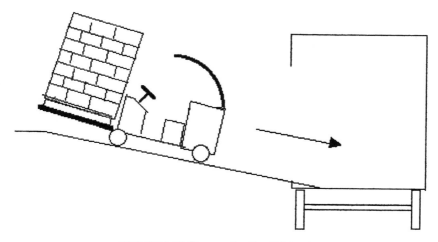

**FIGURE 5.15** Reverse loading lift truck.

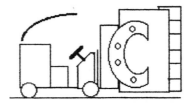

**FIGURE 5.16**  Clamp truck.

warehousemen previously required to store finished production and load an average of three freight cars per day. Although new equipment was required (clamp trucks, live roller-floored freight elevator, and clamp load unitizer), the cost savings on labor paid for the entire system change in less than 9 months. Additional savings were realized from elimination of purchase, storage, and repair of wooden pallets. A

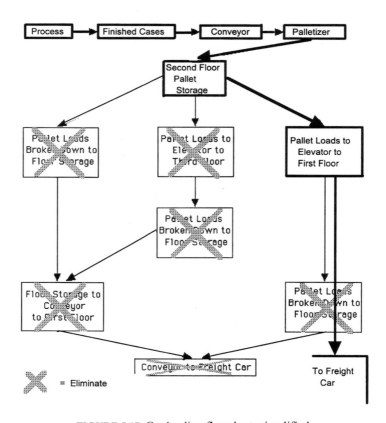

**FIGURE 5.17**  Car loading flowchart: simplified.

surprise bonus was a remarkable reduction in case damage from warehouse handling and car loading.

Relationships between team members were somewhat strained at the outset, but, even without formal training, the antagonisms drifted away as the group concentrated more on the project and less on personalities. The shop steward soon realized that the program of process improvement was inevitable but was assured that the other team members would support his need to retain surplus labor through attrition and by transfer to other positions in the company. The warehouseman, although initially overwhelmed by the "brass" on the team, soon joined in with enthusiasm and turned out to be a particularly valuable member because he had detailed first-hand information of the existing system and was proud of his ability to "fill in the gaps" of practical experience lacking among the other members. The operation manager's assistant provided dozens of ideas, many of which generated more humor than practical application, but, with his comedic approach, he created a friendly and warm atmosphere that encouraged others to make suggestions without fear of ridicule. The engineer provided the technical information required at each step of the discussions and had a useful understanding of the existence and limitations of available material handling equipment. His scientific contribution tended to maintain an even keel to the discussions without stifling generation of imaginative ideas. All in all, the balanced personalities and specialties of the team members were undoubtedly the main reasons for the rapid and innovative solution to this challenging project.

## *Troubleshooting Spice Container Production Variation*

In the production of plastic spice containers, Quality Control noted that the defect rate varied markedly and with no particular cycle. A team was created consisting of the line supervisor, a quality control technician, a line operator, and two plant people from other departments. At the first meeting, the problem was defined as far as possible: reduce defects to some consistently steady acceptable rate. At subsequent meetings, quality records were studied with the intention of creating a Pareto chart, but the inconsistencies from week to week suggested that a Pareto chart would be misleading. It was decided to explore each day's quality problems separately. Following is a transcript of some of the discussions:

Quentin (from Quality Control): July 19th showed abnormally low label defects and no lid damage. We usually get a couple of cracked threads a day. How come?

Oscar (line operator): I don't remember.

Len (line supervisor): I was at a meeting in Toledo that day.

Quentin: OK then, how about July 23rd? Lots of missing seals and thread problems that day.

Oscar: The morning was pretty good. Something went wrong in the afternoon.

Paul (plant operator): What went wrong?

Oscar: I don't remember.

Len: I was in the boardroom in the morning. In the afternoon, I spent a lot of time on the floor speeding up production.

Quentin: The next day, July 24th, was great. How come?

Oscar: Can't remember.

Len: I was out sick that day. Had a severe headache, I think.

A few meetings later, Len was unable to attend.

Quentin: I brought in a whole new set of records today. I noticed that, on some of the days when production rate was highest, the defect rate skyrocketed, but for different reasons each time. What does that mean to you folks?

Paul: Maybe the machinery is too old and can't run any faster unless the preventative maintenance program is improved.

Oscar: I didn't want to mention this in front of my boss, Len, but, every time we speeded up the line, we made a lot of mistakes and the defect bin got pretty full. He's always trying to get us to run faster, but the line can't handle it.

Paul: Tell you something else. Now I don't know much about the line, even after you guys showed it to me, but what I have noticed from the production records and from what we have been talking about is, that every time Len is not there, the defect rate is low, and every time he shows up, he tries to speed up the line, and the defect rate jumps.

Oscar: You're right. Len scares the hell out of us when he's on the floor, and we make a lot of mistakes.

By the end of the discussion, the team decided to break the news to Len as tactfully as possible at the next scheduled team meeting. They created a chart relating line speed to increased defect rate for as many days as they found supporting data. They then agreed to suggest to Len that perhaps the line should be run at a moderate speed until such time as either the engineers could find a way to smooth out production at higher rates or perhaps could suggest improved line maintenance to allow higher speeds. They decided not to mention the increasingly obvious relation between the appearance of the boss on the production floor, attempts at higher production, and resulting increase of defects.

As it turned out, it worked!

## Another Success Story

A manufacturer of top-of-the-line electronic testing equipment created a team to study the unbelievably high rate of rejections at the final test station. The company had recently installed a new quality control system wherein the quality control inspectors were removed from the production lines, and each employee was instructed to be his own quality control technician. As it turned out, the line employees tried hard but were not sufficiently trained in quality control principles. Worse still, they all came to realize that, since there was still going to be a final test station where defects could be found, there was no real need to produce perfect work. The team was hesitant to study this obvious aspect of the problem because of the probability of the highly sensitive political reaction—after all, top management had installed this quality system. They decided to make self-inspection a later project.

Why were so many mistakes being made? The brainstorming session suggested that perhaps the work instructions and specifications might have been written with omissions or mistakes, and this possibility was selected as the first project. By carefully observing the process, the team soon found out that Department A was using specification #1808 modification F. Department B, which received the output of Department A as its starting point, was operating with Specification #1808 modification AD—or 24 revisions more current! The two specifications were incompatible, and, as a result, the final product was riddled with defects.

By distributing the identical current specifications to all departments on the lines, the rejection rate soon plummeted. After this accomplishment, the team had gained enough confidence to consider tackling the sensitive question of restoring the quality control inspection system.

## One More Success

A circuit board gold-plating department was intentionally running the process on the "high side" to make sure everything passed inspection. A team estimated that $1,500,000 could be saved yearly by operating within the specification. By bringing in a quality engineer, the team was able to establish that the department was running out-of-limits on the high side nearly 30% of the time. The engineer also showed the team how the process capability was well within the statistical control limits, therefore permitting reduction of the average (target) plating thickness to the specification average value without running the risk of "thin spots." The process average was adjusted in small steps without any problem. At the end of six months, the process average was set at the optimum level, and, indeed, the new operation saved the company an annualized rate close to the $1,500,000 estimate.

## TEAM FAILURES

The techniques of team formation and operation were originally trial and error, and some of the errors became apparent only when a team project was completed and formally presented to management. In addition to the embarrassment, early failures created an atmosphere of doom for this method. Companies do not talk to the outside world about their team having produced a process improvement suggestion that didn't work. It's embarrassing, it affects employee morale, it becomes a barrier to any future team effectiveness, and it would probably have a disastrous effect on sales. Suppose, for example, a team studies the method currently used to attach each of two hoses to their air conditioner compressor and evaporator intakes and devises a concentric double pipe that carries both high- and low-pressure refrigerant. Specially designed fittings at each end would now permit the hoses to be attached to the air conditioner equipment with only two operations, instead of four.

Unfortunately, when the prototype is first tested, there is an explosion at one of the fittings, causing serious injury and equipment damage. Every effort would be taken to insure that word of this team failure never reached the ears of competitors or customers, and there is no question that the management would make certain to include competent engineers on future teams—if any.

## *A Team Failure Example*

The following example demonstrates another early failure.

During the 1980s, when the team concept was being promoted as a new tool for management, a major manufacturer assembled a team to explore methods for reducing costs of manufacture. Their very first project claimed savings of 2.5 million dollars by modifying a building to house any new missile contract, avoiding the necessity of constructing a complete specialized facility. The modification consisted of extending the roof line of an existing building and constructing three lightweight enclosure walls. Unknown to the team, management had already planned to construct a missile at a plant on the other side of the country in a facility already built and awaiting finalization of the contract. In retrospect, the team was not properly trained in the procedures for assembling all of the data available, for exploring alternative solutions, or for discussing progress reports with affected departments. Subsequent projects tackled by this company's teams were increasingly more successful as the selection and solution techniques were refined.

## *Another Team Failure: Turf Protection*

A newly formed team completed a brainstorming and prioritization procedure and selected a product assembly modification project with great potential for cost saving. When the team approached the department manager to discuss arrangements for scheduling a test run, the manager told the team that "this is *my* department, and no goddamn bunch of hourly workers is going to mess with it. Find some other department to harass. If anybody is going to make improvements around here, it's gonna be *me*." There's not much question what went wrong here. The team concept of process and quality improvement had not been properly introduced to department managers by upper management. In addition, if the team had the slightest suspicion that there might be resistance from the department manager, they certainly should have laid the groundwork much earlier.

## *A Near Failure*

The following example is of a far less serious nature but illustrates a type of failure that can result from insufficient technical knowledge of

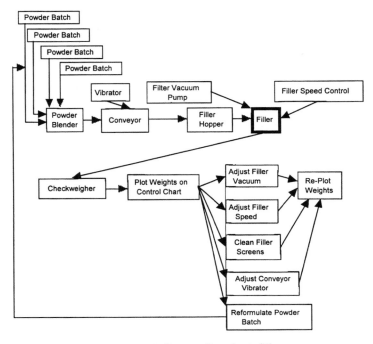

**FIGURE 5.18** Process flowchart: filler.

team members. A team was formed to investigate net weight control of flavored rice mix packages. In any high-speed packaging process where the product is sold by weight, whether the product be cotter pins, paper clips, cocoa mix, or patching plaster, inadequate fill control can eat into profits.

As their first step, the team chose to define the weighing and filling process and eventually constructed the flowchart (Figure 5.18). To their surprise, they found that there was no "scale" device at the filler but that the weight delivered to the packages at the filler was determined by a number of other factors.

They found that periodic samples were pulled from the filler by the checkweigher who determined the net weights of the samples and plotted them on a control chart. When the weight of the sample was found outside of the control limit, the checkweigher had a number of options for correction. She could select one of the following from Table 5.3, depending on her experience and judgment.

The team expressed surprise that the success or failure of weight control rested on the ability of the checkweigher. Their first reaction was to assemble and compare the checkweight records from each of

*Table 5.3.*

|  | To Increase Weight | To Decrease Weight |
|---|---|---|
| Change filler vacuum | higher | lower |
| Adjust filler speed | slower | faster |
| Clean filler screens* | clean | N/A |
| Adjust conveyor vibrator | increase | decrease |
| Request batch reformulation | increase density | decrease density |

*Applied to any blocked filler spout. Mechanic assistance required.

the operators over the previous three months. They suspected that there would be considerable variation in performance but found no consistent differences. The control charts all seemed to be within limits, except for an occasional reading. Generally, the out-of-limits reading was accompanied by a note from the checkweigher regarding what corrective action she had taken, followed by a recheck that generally showed improvement.

Having no success in this direction, the team worked to redefine its mission. In place of "to investigate net weight control of flavored rice mix," they concluded that a more precise project would be "to reduce the spread of the net weight control limits."

As a starting point, additional checkweight records spanning the previous six months were examined. They were hoping to uncover a series of a dozen or so consecutive entries, all of the same value. The intent here was to find the best possible weight record that the present system could produce, to determine the cause for this "best performance," and then to rebuild the system so that it could produce at this quality level continuously. To their disappointment, the only day that showed a series of successive identical weights was five months previously, when every weight was out of limits on the low side for a couple of hours, no matter what adjustments were tried. The problem was found to be an unusual inventory of bulk flavored rice mix with an abnormally light density and no heavy-density material available to blend with it.

Still convinced that there must be a way to tighten control limits, a team member suggested more frequent checkweigh sampling to allow more frequent weight corrections. The line supervisor explained that this would require an additional checkweigher and would be unlikely to show any improvement since the control charts already showed the process was in control. He then stated (correctly) that any improvement would have to be a system change.

Following the suggestion for a process change, the team now proceeded to collect volumes of information on all types of packaging equipment, hoping to find the magic device that would satisfy their goal. They found it, so they thought: an automatic checkweigh machine that would weigh not just samples, but each and every package of mix, with no additional labor required. In fact, maybe the checkweigh person might be eliminated if the automatic machine could be equipped with a series of interfaces that would adjust the filler continuously. In their excitement of discovery, dozens of ideas were proposed: "We could do this . . ."; "We could do that . . ."; "We could attach green and red lights signifying . . ."; "We could send a real-time printout to the blending department"; "We could automatically reject any out-of-limits packages"; "We could eventually add a tare weigher ahead of the filler to get even more precise weights"; and on and on. . . . Unfortunately, this excessive enthusiasm was caused by the weeks of frustration that preceded it rather than the discovery of a solution.

Six weeks later, the automatic checkweigh machine was installed with feedback to the filler to adjust vacuum and speed. It was agreed by all that the expense of inventing integrated controls for the conveyor vibrator and reformulation system might be delayed until the early success of the simpler system was demonstrated. As a safeguard, an ejector was also purchased so that packages beyond the present control limits could be removed automatically from the line.

Figure 5.19 shows an example of a control chart (prepared by quality control personnel on loan to the team) for an average day under manual control and the unbelievably disastrous first day's chart using the new automatic equipment.

What went wrong? The entire team had visions of pink slips and unemployment lines. A number of things seemed to be faulty.

1. The checkweigh machine was generating a control chart for individuals, which has control limits approximately $\sqrt{2}$ larger than the control limits for averages. If a quality control representative had been on the team, he might have explained this to alleviate the fear.

2. Many out-of-limits readings were generated. However, the "out-of-limits" packages were eliminated at the ejector, and although there was an added cost of reprocessing, the material going out the factory door actually contained fewer over- and under-filled jars than normally occurred under manual control.

3. The new machine was overcorrecting. If a package was beyond the

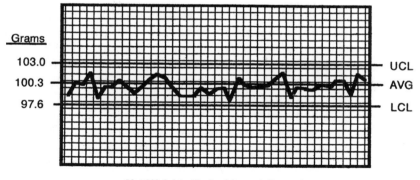

Net Weights Under Manual Control

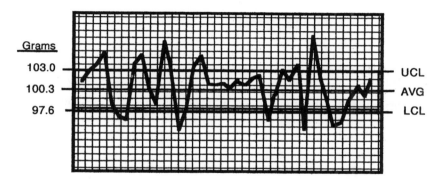

Net Weights Under Automatic Control

**FIGURE 5.19.**

upper limit, the machine would send a signal (for example) to reduce the speed of the filler. If the next package were also beyond the upper limit, another adjustment signal would be sent to the filler, sending the weight tumbling, perhaps even below the lower limit. Then the reverse would occur.

How could this be corrected? After all, there are many automatic checkweighers used successfully. Assuming the filler contains 32 heads, and the line between the filler and the checkweigher holds 40 packages, then there would be 72 packages in the system prior to the weighing operation. This would indicate that, if a package is filled outside of the control limits (for individuals), no further samples should be checkweighed until at least 72 packages have passed the check-weigher following an adjustment. Even this is too short an interval. More time is required for the adjustment to take effect—perhaps another 100 packages. Experiments conducted on the line would soon determine the correct interval for sampling. Once this difficulty was resolved, weight control was greatly improved.

Before leaving the checkweigher project, three more situations should be considered: (1) What happens when the automatic check-weigh machine breaks down? Can the line still be run manually? (2) If the machine attempts to make all of the adjustments in its memory bank without success, can the machine operator or the supervisor override it or perhaps disconnect it while a new batch is blended to correct the problem? (3) If the checkweigher slips out of calibration, how will anyone know? These are complications that the team should consider at the next brainstorming session.

## A TEAM WITH A LONG HISTORY

The concept of teams is not new. McCormick and Company, Inc. established a team known as the "Junior Board" in November 1932. It has been continuously modified over the years and is still in existence under its newest title: Multiple Management. As originally structured, the Junior Board met periodically to consider improvements. Any recommendation they evolved was sent on to the senior corporate board for consideration at their next meeting. One of the unusual features of this early group was that any recommendation *unanimously* passed by the Junior Board would automatically be accepted by the senior board. During the first five years of the Junior Board's activities approximately 2,000 suggestions were accepted. Later, the suggestions were responsible for increasing plant production by 30%.

The Multiple Management system today has a twofold purpose: it provides the means for systematically tackling problems and opportunities and it also serves as a training program for young executives on the way up the management ladder. There are over a dozen Multiple Management Boards, with one for each major unit of the McCormick Company. The size is flexible, but does not exceed 20 members. They usually meet every other week for two or three hours but spend a number of hours of their own time working out details of a project. Members receive extra pay and vacation time. Over 80% of project recommendations have been accepted by senior management.

A few of the many successful projects initiated by these boards were:

- recommended and installed a computerized management control system
- initiated a shipping container design change at savings in excess of $40,000 per year
- designed a shipping and display package at a savings of $37,000 per year
- studied a reorganization of materials management, which was subsequently adopted
- developed a more acceptable procedure for transferring employees between divisions
- provided a system for integrating additions to the company
- devised a procedure for ingredient blending, with significant savings

## SUMMARY

It is unrealistic to expect any single employee to have the wisdom, experience, and training to solve all quality and productivity problems and to have the ability to single-handedly create a constant flow of better quality and productivity concepts. Carefully selected and trained employee teams can effectively contribute to the improvement of every company's operation and profitability. Teams should be a continuing function available to any quality control system.

A constantly supportive management is one of the keys to team success. Support should include providing goals, budget, authority, empowerment, and reward. Assistance from company specialists should be authorized and encouraged by management.

Although some quality/productivity teams require specialized training, all teams should be instructed in nonstatistical techniques: behavior, brainstorming, process flowcharting, cause-and-effect chart construction, Pareto analysis, preparing and interpreting scatter diagrams, and the Deming Cycle. Training should also be available for interpretation of quality control charts, even though the details of preparation of these charts might not be required.

Team goals might be directed exclusively toward problem solving (improvement of quality, process, or administrative procedures) or toward opportunity exploration (new products, services, processes, or equipment).

# CHAPTER 6

# Computerized Quality Solutions and Risks: Statistical Methods

## INTRODUCTION AND HISTORY

A computer salesman tried to convince a manager that purchase of a computer would do half of his work; the manager responded with, "Good! I'll take two." It doesn't work quite *that* well, but the universal acceptance of the computer has greatly simplified the recordkeeping, calculations, charting, and reporting associated with quality control, process control, product improvement, and design of experiments. Considerable knowledge and effort are needed to make a computer perform usefully. A computer will generate beautiful charts, diagrams, paragraphs, or pictures from whatever instructions it receives. If one should blindly feed a computer a list of assorted voltages, wavelengths, and temperatures and then ask it to convert these data into a control chart for air flows, it will probably do just that. Meaningless, stupid, unbelievable? If this hasn't been demonstrated in your company, try it.

Long before computers had been successfully used in quality control, many processes had progressed through a series of quality system development stages. Some companies with adequate staffing were able to proceed through these steps in the laboratory or the pilot plant before actual production began; others stumbled through them on the shop floor.

### Process with No Control

The manufacture of many products starts with no formal quality control procedures, particularly in smaller companies. Each completed item receives a cursory look by the workers, and some of the gross defectives are discarded (Figure 6.1). After sale and delivery, each product is closely examined by the consumer. Quality failures at this late

Production Line

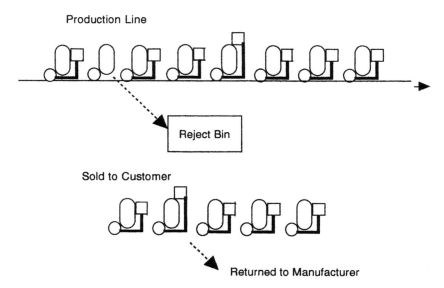

Reject Bin

Sold to Customer

Returned to Manufacturer

**FIGURE 6.1** Process with no control.

stage are usually reported to the seller, who may pass the word along to the manufacturer for action. Depending on the product, "action" might mean return of the purchase price, repair, or replacement of the product. Only after a large number of rejects are noted will any real action be taken.

### Process with Intuitive Control

After a process has been in operation for awhile, some of the workers find out that, if they drop the subassembly on the floor, it probably will cause the finished product to be defective, or, if all of the screws are not inserted, the housing will wobble. Rather than incur the wrath of their supervisor, these workers will control these aspects of their operation. Many intuitive controls arise from casual observations, folklore, and superstition. "We usually have to run the line slower when it's raining out." "Tighten the bolts until the metal first starts to squeak." "If it rattles, tab F is probably not engaged in slot G."

### Investigation with X/Y Scatter Diagrams

This elementary analysis is extremely simple to perform and is basic to any control program. Briefly, it seeks to find if there is a relationship between two factors. Is there a quality connection between line speed and weather? What is the link between torque measurement and bolt squeak (see Figure 6.2)? How many rattle defects are caused by slot misengagement, how many by various lengths of tabs, by thickness of metal, . . . ?

## Sequential X/T Diagram Investigation

If one of the scatter diagrams should indicate a relationship between two variables (rattles and tab length, for example), a carefully super-vised program can be instituted to prepare a chart of all future occur-rences in order of appearance. From this study, it is likely that several system-caused defects will appear. Some possibilities are that, when-ever raw materials from supplier B appear, defects increase (see Figure

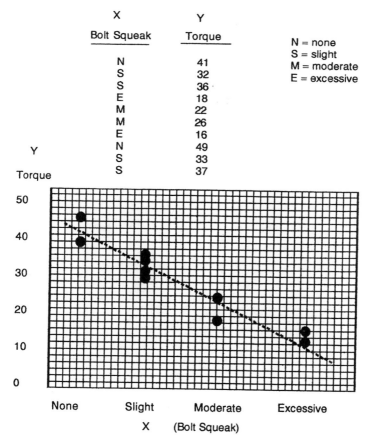

FIGURE 6.2 X/Y scatter diagram, showing relation between torque and bolt squeak.

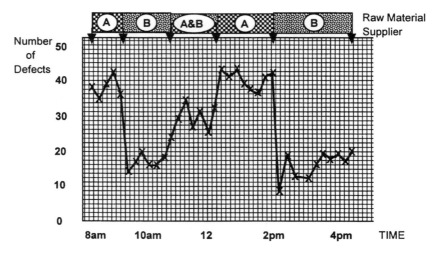

**FIGURE 6.3** X/T sequential chart showing most defects caused by supplier B.

6.3); whenever an inserter machine has just been serviced, defects decrease; whenever Charlie is on the line, defects increase (see Figure 6.4).

- *Process Operated with Statistical Quality Control X-Bar/R Charts*

    Having discovered correctable process variables from the sequential

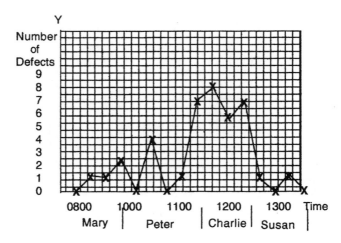

**FIGURE 6.4** X/T sequential chart showing defects over time and identifying employee performances.

diagram study, modifications are made to eliminate them, and a statistical analysis of the improved process is now prepared. Using 3-sigma limits for action limits, it is likely that two benefits will appear:

1. Output will increase, due to a reduction in unnecessary adjustments for imagined faulty conditions.
2. Strong hints of major problem areas will appear, prompting a logical approach to process improvement.

## Product and Process Improvement Studies

Now that the process is under control, further studies may be conducted. Foremost among them would be process capability analysis ($C_p$ and $C_{pk}$) to determine if the process can meet customer quality requirements. Others include statistically designed experiments, evolutionary operations (EVOP), and single factor studies. If a detailed analysis of sample size and frequency was not performed when installing the control chart, now is the time to insure that the sampling procedure is optimal.

## Other Studies

Descriptions of dozens of product and process improvement techniques can be found in books, journals, college classes, and seminars. However, unless a process has been studied with the above stepwise progression in mind to arrive at a state of statistical control, it is unlikely that these advanced techniques will be of much help.

Any qualified quality manager will read the above seven-step history as both obvious and elementary. It admittedly is both. On the other hand, it should serve as reminder of where a present quality control system originated and might indicate whether the current system is based on obsolete premises. For example, suppose that a rattle–defect attribute control chart had been originally designed around misfitting tabs but that tabs are no longer part of the design, having been replaced with spot welds. If this were the case, the sample size and defect control limits might lead to incorrect and misleading information. Reject rates or line stoppages might be inappropriate. Companies who have not reviewed the source of their quality control systems for several years should take a look at them again.

Unless the volume of data is very large, there is no particular advantage to using a computer to prepare the charts illustrated in Figures 6.3 and 6.4 (stages 3 and 4). However, as the data expand and as the calculations become more laborious (stages 5 and 6), the computer becomes an essential tool for quality control and process and product improvement.

## COMPUTER PROGRAMS

Over 500 quality control software companies offer a wide variety of quality/productivity–related programs, and the number increases yearly. Each company offers between one and six such programs, making a staggering total of 1,500 programs from which to make a selection. In addition to this confusing marketplace, 32 companies offer additional specialized programs devoted to development and maintenance of ISO 9000 quality systems, Baldridge Award, and Shingo Prize requirements. Prices for programs vary widely from less than $50 to several hundred thousand dollars. Computer equipment requirements likewise vary from a simple PC to a high-capacity mainframe system. Some of the program features are listed alphabetically below. Items shown in **bold** type are common to practically all basic quality control organizations.

| | |
|---|---|
| Acceptable quality limit | Continuous quality improvement |
| Accuracy graphing | Confidence limits |
| Analysis of variance | Contour plotter |
| Analysis of means | Correlation |
| **Attribute charts** | Cost of quality |
| Average outgoing quality | Criticality analysis |
| Baldridge award criteria | Matrices |
| Bar code access | Mean time between failure |
| Bar code generator | Median charts |
| Benchmarking | Moving average/range charting |
| Box plot | Non-normal distributions |
| Brainstorm format | Normality test |
| Calibration | Operating characteristic curves |
| **Cause-and-effect diagram** | **Pareto** |
| **Check sheet** | Pearson curve fit |
| Chromatographic analysis | Probability |
| Complaint management | Process capability |

Process flow diagram
Process improvement
Quality manual format
Quincunx demo
Real time charting
Regression analysis
Reliability functions
Run chart
Sampling plans
**Scatter diagram**
Scrap and defect tracking
Short run control
Signal/noise ratio
Customer survey analysis
Cusum
Data acquisition
Defect code management
Defect cost calculation
Design of experiments
Distribution analysis
Document control
Failure mode effect analysis
Fault tree analysis
Fishbone charting
**Flowcharting**
Forms design

Fourier transformation
Fractional factorial designs
GMP training and control
Gage control
Gage reliability and repeatability
Gantt charting
**Histogram**
Hypothesis test
Ingredient tracking
ISO 9000 criteria
Simulation
Spectrographic analysis
Standard deviation
Statistical methods
Statistical process control
Stratification analysis
Supplier quality assurance
$t$-test of means
Taguchi analysis
Thermal analysis
Three-dimension plotter
Traceability
Training
Trend analysis
**Variables charts**
Weibull curve

## THE SEVEN BASIC TOOLS

Chapter 4 pictured a quality/productivity system based on three building blocks (Figure 6.5).

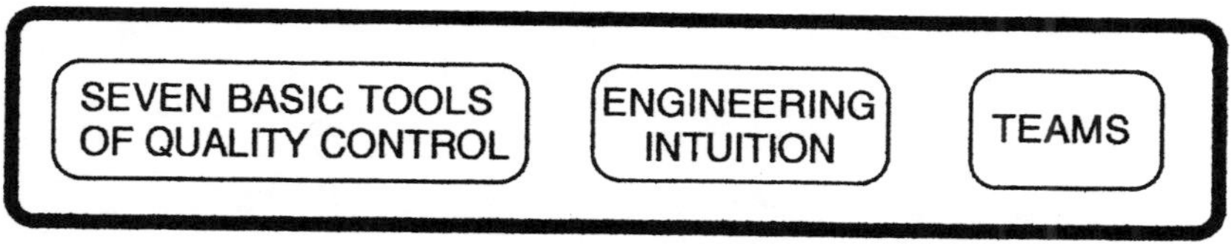

**FIGURE 6.5.**

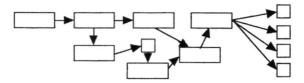

FIGURE 6.6 Flowchart.

So far, most of the seven basic tools have been referred to individually but not clearly identified as a group. Considering the huge number of mathematical and statistical tools available, it is remarkable that so many quality control practitioners agree that an effective system must include the following seven:

1. Flowchart (Figure 6.6)—a picture of a process, using engineering symbols, pictures, or block diagrams, that indicates the main steps of a process

2. Cause-and-effect diagram (Figure 6.7)—a pictorial representation of the main inputs to a process, problem or goal, which detailed subfeatures attached to each of the main inputs, also referred to as Ishikawa or fishbone diagrams

3. Control chart (variable and attribute) (Figure 6.8)—a graph of a some process characteristic plotted in sequence, which includes the calculated process mean and statistical control limits

4. Histogram (Figure 6.9)—a diagram of the frequency distribution of a set of data observed in a process; the data are not plotted in sequence but are placed in the appropriate cells (or intervals) to construct a bar chart

5. Check sheet (Figure 6.10)—generally in the form of a data sheet, used to display how often specific problems occur

6. Pareto charts (Figure 6.11)—a bar chart illustrating causes of defects, arranged in decreasing order; superimposed is a cumulative line chart indicating the cumulative percentages of these defects

7. Scatter diagrams (Figure 6.12)—a collection of sets of data that at-

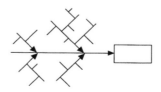

FIGURE 6.7 Cause-and-effect diagram.

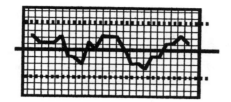

**FIGURE 6.8** Control chart.

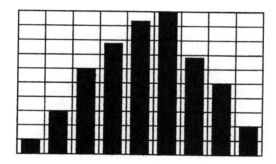

**FIGURE 6.9** Histogram.

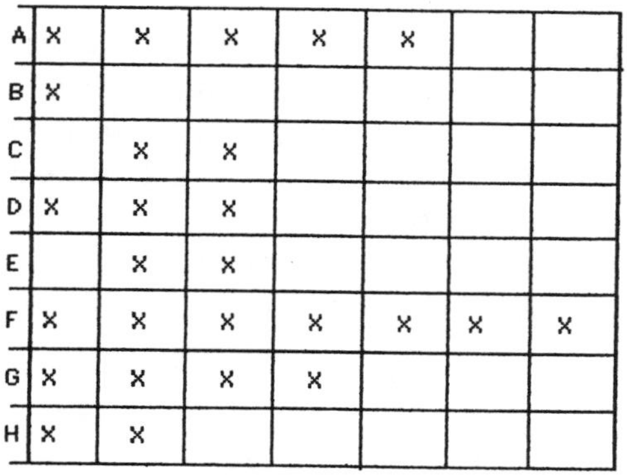

| | | | | | | | |
|---|---|---|---|---|---|---|---|
| A | X | X | X | X | X | | |
| B | X | | | | | | |
| C | | X | X | | | | |
| D | X | X | X | | | | |
| E | | X | X | | | | |
| F | X | X | X | X | X | X | X |
| G | X | X | X | X | | | |
| H | X | X | | | | | |

**FIGURE 6.10** Check sheet.

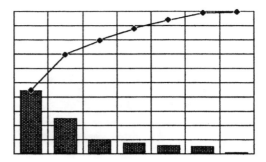

**FIGURE 6.11** Pareto chart.

tempt to relate a potential cause (x-axis) with an effect (y-axis); data are collected in pairs at random

Once these seven basic tools have been built into a quality control system, installation of a computer becomes desirable. "Built into" implies that at least one person in the organization has been trained in the methods of preparing and interpreting all of the tools and has applied them to some of the critical areas of the organization. Those impatient companies that prematurely install a computer program with the expectation of eventually learning how the seven basic tools work are risking the possibility of serious and costly errors, some of which are discussed later.

It is unlikely that one software package will be able to handle all of the potential studies and problems of an individual company. There are no 100% solutions off the shelf. Each application may require a customized program. For example, in some shops, the workers may not be capable of handling mathematics or they may speak English poorly or not at all. They may require an instrument programmed clearly to tell them if something they are working with is good or bad, or, in some instances, they may require some relatively basic control charts to do their job. But, at the middle management level, it may be necessary to perform some serious statistical heavy-duty, number-crunching analy-

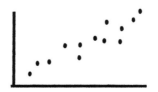

**FIGURE 6.12** Scatter diagram.

sis to evaluate constantly changing problems. Some service companies may need to work primarily with large spreadsheet programs and attribute analysis exclusively. Conversely, some process companies may have no need for spreadsheets and will require only variable analysis. One program will not fit all.

An ideal program coupled with an ideal electronic data collection device would allow the worker to view sufficient data to efficiently operate his process; would also collect sufficient additional data to permit complex statistical analyses, and could graph and otherwise summarize the studies to provide concise management reports. Presumably, the one goal for this system is to provide continuous improvement to both process and quality. The equipment may be new, but the concept is not. The need to enhance the value of a product or service has been with mankind for thousands of years, and it does not imply the need for gigantic leaps forward; constant slow improvement is the safe mechanism of company success.

ISO 9000 requires quality recordkeeping in considerable detail. So, in addition to the information required to operate a production line, perform statistical studies to improve the process, and to report progress and status to top management, the ideal computer program must also provide a structured record-keeping capability.

Some of the special computer program capabilities that might be considered:

- Bring pictures to the shop floor to show people how to do their job. They can be produced in color, in motion, with voice.
- Replace all the notebook instructions on the shop floor. Set up computer displays for documents, specifications, and procedures.
- Describe laboratory procedures, complete with photographs, motion pictures, and report forms.
- Illustrate step-by-step procedure for ringing up a sale, accepting returned merchandise, entering a bank deposit, adjusting a sales slip for discount coupons, viewing product pricing, or checking inventory.
- Formalize maintenance procedures of individual machines, complete with photographs and part numbers, with touchscreen or light pen to permit immediate enlargement of portions of the machinery.
- Diagrams and drawings may be scanned directly from the machinery manufacturer's catalogs for use in these procedures.

The major advantage to these computer aids is that the worker does not have to wander around the office or shop floor to get the information required to do the job; the information is brought directly on the computer screen as needed. Furthermore, the difficulties associated with locating the right person to get information are eliminated: distance, on vacation, coffee breaks, too busy, out sick for the day, personality conflicts, etc.

Common programs, such as Windows, have vastly simplified statistical quality control (SQC) calculations and reporting. For example, if a customer requires a complete quality analysis of a manufactured lot from a vendor, all that may be required is to transfer the spreadsheet production data of that lot to an inexpensive SQC program and let it calculate all of the quality parameters the customer desires: histograms, X-bar and R charts, $C_p$, $C_{pk}$, etc. and then print it on the vendor's report form to accompany the shipment. Complete programs need not necessarily be purchased at the outset of a new computer system; many programs can be purchased as modules, allowing the user to start slowly with the most important parameters. Once these have been formalized, another module can be added. In comparison to employee training costs, computer programs are fairly inexpensive.

## RISKS

Perhaps the greatest risk in depending upon computers for quality control is their ability to produce huge quantities of neatly constructed, very precisely calculated volumes and volumes of charts, graphs, and reports that either have little practical application to the quality control efforts of the company or that are so voluminous that nobody in the organization has the time or inclination to review or interpret them. Of equal concern is that those most involved with quality control reports may not have received sufficient training to understand their meaning. Another risk is that the individual or group responsible for generation of computer applications becomes so engrossed in the power and variety of studies available that complex, accurate exercises in futility are generated, which, if carried to completion, may contribute to the cost of operation without a corresponding process, service, or product improvement.

Use of computers will greatly simplify the calculations and graphing required in the seven-step sequence of procedures (above). But to rely on a computer without reviewing and understanding the source of the system evaluated is a potential obstacle, and perhaps a costly one. The greatest advantages of the computer are the speed with which it

- calculates mathematical formulas
- stores masses of data
- draws charts, graphs, pictures, forms, and plans
- writes
- checks spelling and syntax
- displays data and instructions

By now, most organizations own computers that are routinely used to replace arithmetic and algebra for quality control calculations of statistical equations and control charts. The temptation is to wander further down this road and attempt to apply other statistical techniques that may be very powerful, but realistically should be used sparingly for limited special applications: Operating Characteristic Curves (O.C. Curves), moving averages, box plots, Latin squares, fractional factorials, analysis of variance, mean charts, geometric tolerancing, Taguchi methods, $t$-tests, correlation matrix, etc.

So, where should the quality control program start to use computer analysis on a routine basis? If the product or service can be measured as variables, by all means, the program should be based on X-bar and R charts, process capability, and perhaps acceptance sampling. If the process is based on attribute measurements, the program should be based on the family of p-charts. Steps beyond this should be taken with caution.

Even though it may be programmed to make certain decisions, the computer is not capable of replacing the human thinking process. Following are some painful experiences that resulted from assuming that the computer doesn't make mistakes, never presents illogical answers, and is always equipped with the correct program for the task at hand.

### Faulty Program

An industrial hardware distributor, specializing in oil and gas drilling parts and supplies, decided to improve the customer service quality by computerizing the inventory control system and the manual invoicing procedures. It was believed that having instant access to the inventory of the thousands of parts stocked, accurately updating prices, along with having an automatic computer-generated reorder procedure, would eliminate the back order delays that inconvenienced their customers, often driving them to competitors. Additional benefits would be the virtual elimination of paperwork and the need for expensive periodic physical inventories.

A popular $200,000 computer program with a complete integrated computer system was purchased and installed, with the assurance from the vendor that over 200 such systems had been thoroughly debugged and were running smoothly. Unfortunately, the hardware distributor was not informed that the 200 successful systems were installed on older and different computer hardware. The new computer hardware offered had not yet been tested for compatibility with the program. The results were disastrous. During busy periods, the computer response time would slow down, and on several occasions, when two computers were attempting to access the same information, both would lock and refuse to run. When they were each shut down and restarted to correct the condition, it later turned out that the information they were seeking had vanished from the central computer. For no apparent reason, the system would occasionally type up invoices at a snail's pace and, to add insult to injury, would produce incorrect charges—as much as half price. When trying to fill a customer's order, a clerk found that there was no stock, even though the computer showed several items were still in the warehouse. After a few such experiences, a physical inventory of the warehouse disclosed that the computer inventory records were chaotic.

At the end of a year of these and other severe problems, the company had lost 47% of its gross sales to competitors and had completely reverted to the old hand-entry system. As the discouraged owner of the business worded it, he had no further need of a computerized system of inventory control because his inventory was now small enough to be counted manually.

In discussions with other companies using the same bug-ridden system, it was found that they all had the same series of disasters. In retrospect, it would have been wise to talk to some of the "satisfied computer customers" before purchasing. It might also have been prudent to run only a portion of the newly installed system, along with the old manual procedure, before adopting it companywide.

### Guessing at Computer Program Application

A high-volume manufacturer of a creamy paste filler called in a consultant to discuss how he might conduct quality control classes among the employees. During the preliminary discussions, the quality control manager outlined his procedures, and it became apparent that he was very unsure of quality control techniques. For example, when asked if charts were in use on the production floor, he displayed a computer-generated "control chart" for individuals, showing three-sigma limits and explained how he took a sample of one paste filler container periodically, calculated the $C_{pk}$, and determined from that if the process was in control.

Apparently, the manager was unaware that $C_{pk}$ can be calculated only *after* the process has been shown to be in control, and only then can it be used to determine how far the centerline of the process has drifted from the optimum. It is *not a* control calculation. Rather than explain the risks associated with misunderstanding the function of $C_{pk}$, the consultant tactfully asked how this control procedure had been developed. The manager then explained how, when he was promoted from a production job to his present position, he found a computer quality control program with many options for graphing data. He selected the $C_{pk}$ as one that "could be easily understood by the line personnel." Unfortunately, this new manager did not fully understand the concept.

### Faulty Management Systems

The Department of Motor Vehicles (DMV) of one of the largest states was attempting to speed up the notoriously slow service in providing licenses, registrations, and other services by installing a massive computer system. The data banks of the DMV covered dozens of entries: name, address, sex, color of hair, height, weight, electronic thumbprint, records of moving violations, parking violations, accidents, and car(s) owned, along with the make, model, color, year, registration number, date of renewal, smog record, and insurance data. There was little doubt that tying the department's computers into one central data bank should sharply reduce delays at the field offices.

Over several years, spanning two governors' terms, the program failed to accomplish the objectives of each stage of its development. In spite of this, the department continued to spend large amounts of money without first correcting the failures. Even the accounting functions of the project were questionable: the total project was based on $44 million to be spent on hardware, software, installation, and consulting services, but this budget was underestimated by over $5 million. Although the software was prepaid, there is some doubt as to whether the complete package actually was received. Finally, the DMV director told the lawmakers that the project had been abandoned.

In a disaster of this magnitude, there is bound to be much finger-pointing regarding the responsibility, and many have been justly accused of mismanagement. In the hope of advancing the project quickly, some fundamental management procedures had been skipped. A contract had been awarded without competitive bidding, some contractors were given permission to start work before the contracts were awarded, and the contract approval organization was bypassed.

Perhaps the basic problem was lack of planning and failure to adhere to such a plan. Certainly, a small-scale working model of the total operation should have been successfully demonstrated prior to embarking on this massive project.

### System Backup

There is general agreement that every computerized system must have a backup copy of the program and the data. There is another very important system backup that is sometimes overlooked: a manual backup system in the event of a computer failure. In the retail business, consider the effect on customers when the clerk says,

- "I can't write up your order; the computer's down."
- "I can't take your money because the computer change calculator isn't working."
- "The scanner isn't operating, and I don't know how much this stuff costs."

Knowing that computers can fail, there is little excuse for not providing an alternate system. Most banks have instructed their tellers to handwrite deposits and withdrawals. Most hospitals have trained their admitting nurses in procedures for writing records for patient admission. Many manufacturers know the dangers in failing to provide instructions for manual operation of computer-controlled machinery. An example of standby procedures in the event of a process computer failure is

When the red signal light on the temperature controller flashes, it indicates that the process is now in manual operation. Use the hand valve located below the temperature gage to raise or lower the temperature as required. Failure to do so will result in the process lines congealing at low temperature or exploding at high temperature. Maintain temperature between the red and black control lines shown on the thermometer, and press the large red panic button to signal the need for immediate assistance.

## Misuse of Computer Program

One of the favorite quality control computer programs provides a large number of statistical charting techniques that can be used to simplify data. It is important, however, to fully understand what these techniques mean. If requested, the program will willingly plot a Pareto curve for anything desired, regardless of what the x-axis signifies. The following example should illustrate how ridiculous the results can be.

A computer technician, untrained in either engineering or quality control, collected data on all the "bad things" that had caused stoppages on the production line over the past three months. He found eight categories that had occurred more than two or three times and fed this data into a computer programmed to produce a Pareto chart. The "bad things" were as follows:

|     |     |
| --- | --- |
| 63  | wrong size ratchet wrenches on the floor |
| 238 | spray cans contained dark green paint instead of the standard |
| 25  | products were severely chipped when they fell off the line |
| 46  | counterweights were too heavy |
| 12  | days the line started up late (materials were not ready) |
| 163 | mornings were fairly dark because of heavy rains |
| 4   | times, Mabel the tally master did not show up for work |
| 34  | occasions during which the computers were down |

By his punching in the instructions to calculate the format to draw a Pareto curve, the computer ranked the occurrences, calculated the percentages, and printed a Pareto chart as shown below in Table 6.1 and Figure 6.13.

The typical 80% cause factor appears (actually 79.32%) after the first three classifications: too green, raining, and wrong size. The figures are impressive and undeniably accurate. The graph looks very scientific. But the only sense that can be gleaned from this study is that the paint supply should be more accurately controlled and the chances are that this was already known. It might be very difficult to find a

*Table 6.1. Bad Things That Happen.*

| Bad Thing | Number | Percent | Cumulative% |
|-----------|--------|---------|-------------|
| Too green | 238 | 40.68 | 40.68 |
| Raining | 163 | 27.86 | 68.65 |
| Wrong size | 63 | 10.77 | 79.32 |
| Heavy weights | 46 | 7.86 | 87.18 |
| Computer down | 34 | 5.81 | 92.99 |
| Chipped | 25 | 4.27 | 97.27 |
| Started late | 12 | 2.05 | 99.32 |
| Mabel not here | 4 | 0.68 | 100.00 |

cause-and-effect relation between production stoppages and the rainy weather, or the presence or absence of Mabel, for that matter. Nowhere is the dollar value of the causes or effects shown, making the entire effort pointless. It is possible that the 238 stoppages for paint problems required only one or two seconds each, for a total of only four minutes during the entire period studied, whereas the chipped products might have cost the company many thousands of dollars from both lost time and scrapped product.

The computer is a remarkable tool that can absorb massive amounts of information and can produce professional-looking tables and graphs. But great care must be exercised when selecting the input data. Greater care must be exercised when interpreting results.

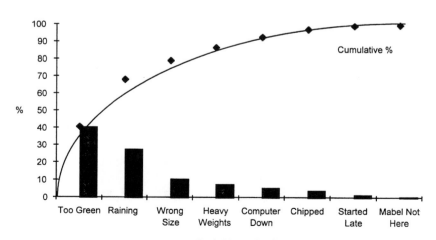

**FIGURE 6.13** Bad things that happen.

## Back up New Program Data

Several years ago, a major professional organization of approximately 80,000 members had relatively little contact with the membership and was able to handle monthly mailings with the use of a fairly high-speed copy machine and secretarial help. As their services grew, however, it became apparent that the secretarial staff was unable to handle the load, and management was convinced of the advantages of computerizing the membership list.

Contracts were drawn up with a major computer company, and the membership names, employers, addresses, phone and fax numbers, grades, specialties, and other data were soon transferred from the office files to the new computer memory bank. A few days after the system was "up and running," it mysteriously crashed, and some unknown amount of the input data was lost. Worse still, the person responsible for the files transfer had discarded all of the original paper records and, because of complete confidence in the new computer system, nobody had thought to create a new printout for reference. It took well over a year to reconstruct the data.

The story of this unfortunate experience, and others just as serious, was widely circulated. Consequently, today (it is hoped) nobody sets up a new system without backup.

## Program Capability

A major computer programmer prepared an application that combined vast amounts of production and distribution data for major product lines of a manufacturer and then calculated the least cost combination. To further complicate the calculation, it recalculated the minimum costs at various quality levels.

The first attempt to run the program with real data proved to be a serious disappointment. The costs reported were ridiculously low and the distribution quantities exceeded the production levels—obviously an impossible condition. Equally impossible, production at all quality levels was reported to incur identical costs. The cause for this confusion was quickly located by the programmer. In an attempt to avoid excessive computing time that would be required to carry out all calculations to 15 or 20 significant figures (production levels were normally in the millions of gallons), the program was created with a 1.0 "rounding factor." As a result, all costs per hour that were fed into the computer to the nearest $0.0001 were rounded off to $1 or zero. This problem was easily corrected by installing a more realistic "rounding factor," and the program was run quite successfully thereafter. On the other hand, to the uninformed, a gnawing question remained: supposing the programmer increased the factor even further—would the results be the same?

Months later, an even more frustrating problem appeared when the program was asked to calculate the least cost to operate the plant during the vacation months of the year. Should there be overtime production before or during the period? What production levels would be required ahead of time if all vacations were taken during the same month? Should a centrally located warehouse be overloaded prior to the vacation period to take up whatever unusual shipping need arose? Should only the medium quality product inventory be expanded, or should the inventory of higher volume market quality product be the one to build up? The program very quickly responded to all of these problem data: the least costly production schedule was to not run at all! Again, the programmer was able to provide a fairly simple solution by commanding the computer to insure that the plants would be operated during the vacation period. Once more, the computer illiterate wondered what else the programmer forgot to tell the program.

To summarize: computer programs are only as useful and reliable as the information designed into them. When used for purposes other than those for which they were intended, they might produce inaccurate—or even dangerous—guidance. Installing a computerized system should be a stepwise procedure, just in case there is a problem in the program, the concept, or the equipment.

## BENEFITS OF COMPUTER

### Computer-Controlled Processes

Over the years, industries have labeled each new set of operational system and control tools with descriptive terms:

| | |
|---|---|
| — | Technocratics |
| — | Technology |
| IE | Industrial Engineering |
| OR | Operations Research |

| MBO | Management by Objectives |
|-----|-------------------------|
| MIS | Management Information System |
| CAPC | Computer-Aided Process Control |
| COMM | Customer-Oriented Manufacturing Management |
| MRP | Materials Requirements Planning |
| MRP | Manufacturing Resources Planning |
| DRP | Distribution Resources Planning |
| ERP | Enterprise Resources Planning |
| EDI | Electronic Data Interchange |
| PMIS | Process Management and Information System |
| CIDC | Computer-Integrated Distributed Controls |
| MOMS | Manufacturing Operations Management System |
| CIM | Computer Integrated Manufacturing (Management) |

For one reason or another, those organizations that have developed computer-assisted operation controls have preferred to invent and use their own nomenclature, but the one term that seems to encompass the intent of the most recently introduced concepts in the above list is CIM, or computer-integrated manufacturing. Even this may have been an unfortunate choice of words, since it seems to exclude the computer-integrated operations (CIO) of nonmanufacturing service organizations. The obvious conflict of these initials with those of a major labor union may be the reason. To be more inclusive, perhaps CIM should stand for computer integrated *management*. It may be somewhat simplistic to classify all of the above-listed items as CIM, but they do all have two characteristic features: they are all computer driven and they are all intended to improve efficiency, profit, and quality.

Computer-integrated systems have been developed to save labor costs, to eliminate human error, to optimize service and processes, to reduce or eliminate variation, and to provide speedy and accurate computations involving huge quantities of data. Many of these goals are the same ones that drive statistical quality control.

Three of the everyday applications of CIM are banks, supermarkets, and fast-food outlets. When a deposit or withdrawal is made at a bank, a few simple computer keyboard entries guide the transaction through the intricate maze of records to the correct file: customer, deposit, withdrawal, loan payment, interest or principal mortgage payment, checking account, savings, money market, certificate of deposit, Christmas fund, or security purchase or redemption. Then it is added, subtracted, or calculated to provide a new balance and is simultaneously transferred to a printed receipt or passbook. In addition, the bank's master files are instantly updated to reflect the increase or decrease in category balance, total daily transaction, and cumulative summaries. Other bookkeeping data may also be entered and compiled: teller identification, date and time, account number, and interest earned or charged. Checks received during the day are bundled and sent off to a central station where they are decoded by a computerized scanner system with the retrieved accounting data returned to the appropriate bank.

Many supermarket computer systems at the checkout counter are programmed to perform a variety of calculations beyond adding up the cost of items purchased. The combined invention of the universal product code (UPC) and the code scanner has presented the supermarket industry with a remarkable opportunity to increase the quality of customer service and business management. Scanning the UPC of discount coupons as well as products has eliminated annoying long waiting lines at the checkout counters. For the most part, accuracy has been dramatically improved, although there are some remaining problems with speedily updating price changes in the system. But the greater advantage to the system is the opportunity to instantaneously transmit checkout counter data to the central office computer to determine hourly inventories (when needed) of every item handled. When coupled with periodic physical inventories (usually conducted with the use of hand-held computers), each store is capable of accurately determining losses from pilferage and damage. In addition to reducing inventory control problems (such as out of stock), the massive amount of data constantly updated permits accurate studies of the advantages of special sale items, the relative sales and profits of competing brands, and the ratios of cost to profit for various store layouts (number of checkout lines, number of aisles, effect of special lighting, operating of a florist service, delicatessen, or bakery in the store, etc.)

Fast-food outlets originally installed computer systems to provide speedy and accurate customer service. It soon became apparent that the computer was capable of expanding the concept to provide input for more efficient control of the inventory, materials supply, labor and other cost accounting, procedures, maintenance, sanitation schedules, and reporting forms—in short, complete outlet system operation. A logical extension to this has been the direct wire collection of operational data from countrywide outlets to the headquarters office, where production and distribution schedules, purchasing requirements, cost analyses, personnel records, and all of the other company operations could be computed, analyzed, and efficiently managed. Some fast-food chains use computerized cash registers that use symbols for the items ordered, thus enabling the clerk to punch a single key for small fries (for example), rather than several keys representing the price for that item. This increases both speed and accuracy. Some cash registers permit the clerk to depress the "total" key (which will include the calculated tax), followed by the amount paid by the customer. Automatically, the correct change is figured and illuminated on a screen. Thus, in the event that a total bill is $2.86 and the customer (not wishing to receive a lot of small change) tenders $3.11, the computer instantly calculates $0.25 change and posts it on the screen, reducing the possibility of error. Considering the prevalence of marginal math skills in some of the work force, this is a decided advantage to both the customer and the management.

In the recent past, the goal of many manufacturing organizations was to automate each department with the installation of mechanical, pneumatic, and electronic controllers. These unit operations became increasingly efficient to the point where unit managers began to consider methods of combining their efforts. The rapid development of computers provided the tools to accomplish this end. When used in these industries, computers have had an unbelievable impact on quality and process improvement and control. It is difficult to generalize on this subject because each company has independently "computerized" its operations according to its own beliefs and needs.

## Electronic Measuring Device Input

Had it not been for the development of instantaneous electronic signal output from control devices, real-time computer control could not exist. Translating the enormous volume of visual, mechanical, or pneumatic signals from a measuring device to the keyboard of a computer would have required a huge delay.

Control devices have had an interesting history. Most of them started out as passive monitors: a thermometer, a pressure gage, a weighing scale, and a volt meter. An equal arm balance, for example, could display weights by mechanically connecting the arm to a pointer rod that terminated on a numbered scale (Figure 6.14). Mechanical modifications could provide a primitive degree of automation by automatically dumping the contents of a hopper on the weighed arm (Fig-

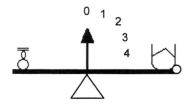

**FIGURE 6.14** Simple dial indicator.

ure 6.15). By connecting the arm to a valve, pneumatic signals could be carried substantial distances to actuate other processes or numerical readouts (Figure 6.16). Pneumatic signals had a number of complications and inaccuracies, and were gradually replaced by electrical signals generated by arm movement (Figure 6.17). The subsequent application of electrical strain gages to the design of weight-measuring instruments introduced electronic dial readouts and, of greater importance, permitted direct input of weight data to a computer (Figure 6.18). By programming the computer to act on this weight information, a vast array of actions could then be initiated instantly.

Similarly, other passive measurements of light intensity, vibration, reflectivity, sound, pressure, specific gases, moisture, foreign objects, distance, porosity, pH, position, curvature, smoothness, colorimetry, and many other product quality attributes have all been converted to active electrical impulses that can now be sent directly to a computer that has been programmed to receive instructions to record, display, control, or analyze.

Rather than depend on individual controllers to guide segments of a process, programs were developed whereby all of the process control instruments were gathered together in a computer for total control. Any sequence of the process could be displayed on the controller's computer screen, as shown in (Figure 6.19).

By selecting icons, the operator can instantly identify all of the cur-

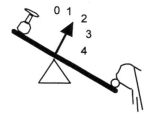

**FIGURE 6.15** Automatic dump.

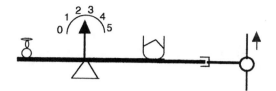

**FIGURE 6.16** Pneumatic valve actuator.

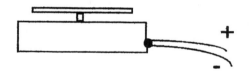

**FIGURE 6.17** Variable resistor output.

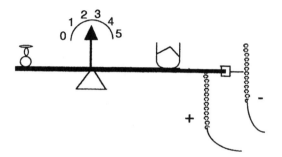

**FIGURE 6.18** Electronic scale.

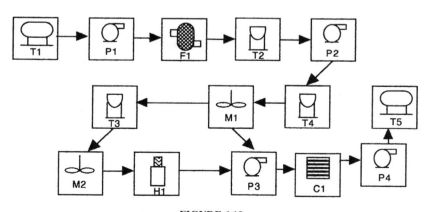

**FIGURE 6.19.**

rent data at that location as well as process specifications for that stage. Process corrections might also be included. For example, in the event that the cooler signaled an abnormal functioning (perhaps by an intermittently flashing C1 icon), the operator would select cooler icon C1, and would find the following on his computer screen:

| COOLER C1 | | | |
|---|---|---|---|
| Time 0837    Date xxxx | Current | Target | Limit |
| Flow | **36.95** | 34.00 | 32-36 |
| Temperature | 68.3 | 70.00 | 68-72 |
| Viscosity | 32 | 32 | 30-34 |
| Pressure | 21 | 25 | 20-30 |

ADJUST: REDUCE COOLER C1 FLOW IN 0.5 STEPS BY CLOSING VALVE AB335 ONE CLICK STOP EVERY 30 SECONDS UNTIL 34 IS REACHED.

The operator can then select a number of quality control analyses of this process, such as a variables control chart for averages and ranges. This would instantly provide a history of flow analyses, showing trends, frequency, and direction of outliers, cycles, and patterns. From this information, a skilled operator would be in a position to suggest maintenance or repair of specific equipment. In the event of a serious problem, the quality engineer (who would also have access to the same computerized information) would be notified for an immediate response.

Some operations, such as the above, have been programmed to continuously monitor the data automatically, signaling the operator when any reading is out of control. An industrial process that might have required two or three operators in the process control room and four or five operators on the floor (if run manually) could now be controlled by a single operator at a computer console, thus using the computer as a miniature process control room.

The power and efficiency of computer-controlled operations can be extended well beyond the production floor. Companies have found that all of the manufacturing support functions can also be linked. Each department still functions as a specialized organization, such as quality control, accounting, engineering, marketing, procurement, etc., but they are tied together by a computer network to function as a multidisciplinary team. The needs and plans of each department are made available to all. Huge, complicated, multioperational companies have managed to link their units together with computer networks, permitting the entire organization to be operated as a single entity.

Without such a computer program available to plan for the following year's production schedule, a company that depends on committee meetings might encounter the following scenario:

- marketing—We expect to sell two million units in the first quarter, five million in the second, three million in the third, and one million in the fourth.
- production scheduling—That forecast will require 60 shifts in the first quarter, 150 in the second, 90 in the third, and 30 in the fourth quarter.
- manufacturing—The first, third, and fourth quarters look OK, but we cannot produce five million in the second except on three shifts per day.
- quality control—The third shift requires seven extra quality personnel.
- accounting—Three shifts will cost too much. Move extra production into the first quarter, and put the excess in an outside warehouse.
- engineering—We can't. Two lines will be down for maintenance in February.
- personnel—Extra shifts in vacation period violate union contract.
- operations—We could farm out extra production to a contract packer.
- quality control—We need to have two of our quality engineers at the packer.
- accounting—No more than one million to the contractor, or else profit margin suffers.

Assuming some kind of compromise production schedule is reached by the committee, a subsequent meeting might start off with

- marketing—Oops! We now find we need three million in the first quarter.
- CEO—How long will it take to install another line—and at what cost?

By contrast, with a manufacturing resources computer program, all of the above data may be fed into the computer files to provide a solution to the lowest cost combinations of shifts, contract packers, and monthly production quantities.

## SUMMARY

We have seen how a computer can be programmed to provide instant access to quality control data, charts, and reports. The vast array of quality control-related computer programs are capable of performing highly complex calculations for studies to improve both quality and process. By networking the quality control computer system with other departments in an organization, unit control can be melded into an integrated and unified system where information is immediately available to all departments. This network provides the shared data for efficient calculations involved with production scheduling, cost accounting, quality control, warehousing requirements, personnel administration, purchasing, maintenance, and all other functions of the organization, thus reducing the need for costly time-consuming staff meetings.

Attempting to install a computerized quality control system before the basics are fully understood carries with it a high risk. Unless someone in the organization is familiar with the basic quality control principles, there is a possibility, along with other risks, that the wrong program might be selected, thus furnishing meaningless or even harmful information. In the event of computer equipment failure, the entire operation can come to a halt unless someone in the company is capable of providing manually calculated controls.

The basics referred to above are (1) the seven quality control tools, (2) a problem-solving procedure for enlisting the engineering and intuitive knowledge of experienced employees, and (3) a formalized system for investigating quality and process improvement through the use of trained teams.

# CHAPTER 7

# Making the Right Choice

MAKING the right choice also applies to individuals. People consciously or unconsciously make decisions on how they control the quality of their lives. Their level of intellectual pursuits in music, philosophy, art, literature, and religion can be characterized. They establish the quality control of their physical well-being, their number and type of friends, their eating habits, their consumption of stimulants such as alcohol or tobacco, their selection and frequency of physical and entertainment activities, and every other facet of their lives. But this does not mean that they are satisfied with either the level or the control of the quality of these many aspects of existence.

Similarly, an analogous situation exists in manufacturing companies, process industries, and service organizations (such as schools, hospitals, banks, retail stores, churches, law offices, consultants, government, transportation, etc.) Regardless of size or age, each organization operates under a chosen principle of control over its output at some level of quality. Again, this does not mean that it is satisfied with its quality control.

If an organization wishes to improve its quality level or control of that level, there are certainly enough techniques available to effect a change. Unfortunately, as was emphasized in the early chapters, the abundant number of choices cannot be sampled indiscriminately. Some guidance is necessary.

It has been suggested in Chapter 4 that all quality control systems should be based on three techniques. If an organization wishes to improve quality control, here is the place to start:

1. At least one employee shall be thoroughly trained in and familiar with the seven basic tools of quality control.
2. Creative individuals with long-time experience, or technically

trained personnel, shall be called upon to assist in the solution of quality system problems.

3. The management shall actively support the training and operation of cross-functional quality improvement teams.

How does a company build a quality control system, once the three essentials have been firmly established? In addition to the extensive literature on this subject, there are two major sources available as guides:

1. The Malcolm Baldridge National Quality Award criteria, especially for the year 1990. This concise outline of requirements offered an excellent base for a quality control system at that time. However, over the years the award criteria have presented a moving target, shifting from *quality* control to *business* control. For example, one of the seven award criteria has been modified as shown below:

|  | Evaluation Grade (%) |
|---|---|
| Item 6.0 Quality Results—1990 Award |  |
| Quality of Products and Services | 5.0 |
| Comparison of Quality Results | 3.5 |
| Business Process, Operational and Support |  |
| Service Quality Improvement | 3.5 |
| Supplier Quality Improvement | 3.0 |
|  | Total 15 |
| Item 6.0 Quality and Operational Results— 1994 Award |  |
| Product and Service Quality Results | 7.0 |
| Company Operational Results | 5.0 |
| Business and Support Service Results | 2.5 |
| Supplier Quality Results | 3.5 |
|  | Total 18 |
| Item 6.0 Business Results—1995 Award |  |
| Product and Service Quality Results | 7.5 |
| Company Operational and Financial Results | 13.0 |
| Supplier Performance Results | 4.5 |
|  | Total 25 |
| Item 6 Process Management—1998 Award |  |
| Management of Product and Service Processes | 6.0 |
| Management of Support Processes | 2.0 |
| Management of Supplier and Partnering Processes | 2.0 |
|  | Total 10 |

As another indication of the shift of emphasis, one might note that in 1990 the award promoted ". . . understanding the requirements for quality excellence." In 1995, the wording was changed to ". . . understanding the requirements for performance excellence." Note how the word "quality" has disappeared from this category.

However, the basic requirements outlined in the 1990 (and earlier) National Quality Award criteria present an excellent framework of goals for a complete quality control system.

2. ISO 9000 Series (also known as ANSI/ASQ Q90-94). Although the major field of application is the manufacturing industry, the standards can be interpreted, with some effort, to apply to service industries as well. These quality management and quality assurance standards are as follows:

- ANSI/ASQ Q90—Quality Management and Quality Assurance Standards—Guidelines for Selection and Use
- ANSI/ASQ Q91—Quality Systems—Model for Quality Assurance in Design/Development, Production, Installation, and Servicing (for use when conformance to specified requirements is to be assured during several stages, which may include design/development, production, installation, and servicing)
- ANSI/ASQ Q92—Quality Systems—Model for Quality Assurance in Production and Installation (for use when conformance to specified requirements is to be assured during production and installation)
- ANSI/ASQ Q93—Quality Systems—Model for Quality Assurance in Final Inspection and Test (for use when conformance to specified requirements is to be assured solely at final inspection and test)
- ANSI/ASQ Q94—Quality Management and Quality System Elements—Guidelines

Although these standards are required to be reviewed every five years, changes are expected to be modest and should not alter the basic structure of the existing documents.

Rather than choosing one of the above two sources as the exclusive format for a quality control system, it might be wise to study several other suggestions and select those portions that are most likely to work for a specific organization. Whatever system is evolved, the framework should be documented so that it can be reviewed conveniently by present and future management. The system need not be static. In fact, it should be remodeled as needed to keep in step with changing market conditions and changing company needs.

In the retail business, people are motivated to buy one product in preference to another because of perceived advantages, frequently as the result of persuasive advertising. Selling quality control within an organization can be approached in much the same way. Employees may be cooperative for various reasons, many of which resemble those selected in retail advertising:

| Reason | Advertising Technique | Selling a Quality Control System |
|---|---|---|
| Authority | We've been offering this product to the public longer than anyone else; we are therefore the best in the business. | I'm the boss; I've been in this business longer than you; so do as I say. |
| Generosity | New product—buy one, get another at 50% off; join now and get a free checking account with interest. | In appreciation for your cooperation you can expect a promotion with an increase in pay. |
| Commitment | Thousands of our customers have been with us for over 20 years. | Dedicate yourself to this job, and you will succeed. |
| Scarcity | We are the only one who has been able to perfect this product. We can hardly keep up with demand, so buy now before it is all gone. | Challenging jobs that pay as much as yours are difficult to find. |
| Friendliness | Our customers are our finest asset. We guarantee that your experience in any of our stores will be a pleasant one. | We consider everyone in our company part of our happy family; we refer to you as an asssociate, not as an employee. |
| Consensus | Surveys have shown that ours is best. | Our entire industry agrees that we are doing things right. |

## MECHANICS OF CHOOSING (OR IMPROVING) QUALITY AND PROCESS

### No Formal Quality Control System

First of all, establish a written quality policy.

Select and train an employee in the seven basic tools of quality control to initiate a quality control system. Hiring an outside consultant at this stage is rarely as effective as utilizing an existing employee who is familiar with the operations of the company. Implement a control chart system on the major product or service lines of the company, and establish reporting routines. Familiarize the entire organization with the power of the control charts generated by the quality control function. Conduct studies, using the basic tools, to improve the quality or the process, and *publicize the dollar savings.*

Next, assign to an engineer or long-term employee (or both) the responsibility for finding solutions to existing quality or process problems. Select and train teams to explore product and process improvement studies, using quality control and engineering as specialized information sources.

Give this organization a year or two to become firmly established before looking to advanced quality techniques.

### Quality Control System in Use, but Antiquated

If the existing system consists of inspect-rework-scrap, it is time to rebuild it. The system should not be abandoned until the new one is in place, and has been running parallel long enough to demonstrate its effectiveness. By following the steps in the previous section, it should be possible to install an effective quality control system without disrupting the business.

If only a portion of the three-prong quality system is in place, the other functions should be added. For example, if the seven basic tools along with competent engineering and intuitive personnel are in place, but team techniques have not been tried yet, it is suggested that they be given a chance before attempting other quality ideas.

### System Revision Needed to Accommodate New Product, Process, or Quality Policy

If an effective basic quality control system is operating satisfactorily, but a major change has been made to the product, quality policy, or process, a new set of procedures might be required: sample size, location, and frequency; method of data collection, charting, analysis, and routine reporting; or action taken and follow-up. Management may not be aware of the time and effort required for all of these extra operations, and it might be advisable to consider planning for additional trained quality personnel during the initial stages of a new operation. In some instances, outside consultants or temporary technical help might be considered. Once the new operation becomes routine, it might be incorporated into the existing system without additional personnel.

### Install Computer or Improve Computer Usage

Computer marketers may promote the speed, accuracy, and wonders of their equipment, but, before they are installed as controllers of a system, that system should first have been demonstrated to perform satisfactorily under manual controls. Much like the passenger in a high-speed automobile whizzing down the road who is unable to observe the weeds and diseased trees alongside, high-speed computer operations may not be instructed to control all of the defects in a system. Once the weaknesses and danger points of a process have been duly noted, understood, and perhaps eliminated, then is the time to consider high-speed computer measurement and control.

Rarely do computer-assisted quality procedures drop into place without complications. Frequently, this is because the awesome power of the computer is not fully considered during the planning stages, and additional valuable functions are discovered during the initial operations. Another class of surprises is due to the lightning speed of computer calculations, which permits or creates instant response time. In some cases, this can result in overcontrol, with its resulting quality degradation. Both of these situations are readily corrected by careful review.

Although the computer is a highly desirable tool for improving control of quality, it still requires that the individuals depending on its output have a thorough knowledge of the basic seven quality tools. The computer input requires knowledge, and the computer output requires interpretation.

Computer operations should be reviewed periodically. New processes might not be controlled satisfactorily with existing input systems or control programs. As the quality level is improved, the old control limits may no longer apply. It might also be advantageous to periodically consider newer and more powerful programs.

## Investigate and Adopt New Techniques

As pointed out previously, it should take a year or two of basic quality control operation before investigating the newer techniques. Calling in a consulting team that promises installation of a Total Quality Management system in a few months has often proven to be a disappointment. Some of these changes require three or more years to perform satisfactorily, and, by that time, processes become obsolete, and people, product, and service may have changed. This is not meant to imply that the newer techniques cannot be advantageous. On the contrary, many have provided outstanding performance improvements.

Periodically, management should review and update the company quality policy. If the company is showing a profit, if all of its operations are satisfactory, and if all of its customers are pleased, adopting new policy techniques would appear to be risky. However, there are few companies, if any, who continuously meet these conditions to the satisfaction of management and the stockholders.

Many of the newer techniques can be added to an effective quality control program without upsetting existing systems. Some that might be considered are

- computer-integrated management
- benchmarking (internal and external)
- statistical design of experiments
- total quality management
- vendor certification
- customer survey

A final word on total quality control. Many have attempted to deal with this concept as a clearly defined, absolute technique and have been disappointed with its failure to perform the expected miracles. This does not suggest that a total quality program should be *flexible* (in the sense of being spineless) but rather that the program should be *adaptable,* seizing on opportunities for continuous improvement. Those organizations that have treated total quality control as a *goal* are far more likely to experience a series of successes.

ASI. 1990. Total Quality Management, Version 4.0, AM. Supplier Institute, Dearborn, MI.

ASQ Food, Drug, Cosmetic Division. 1998. Food Processing Industry Quality System Guidelines.

Bartlett, P. N., Eliott, J., Gardner, J. 1997. Electronic noses in the food industry. *Food Technology*, 51(12):44.

Becker, S. W. 1993. TQM does work. *Management Review*, March 1993.

Bentowsaki, K. 1992. The quality glossary. *Quality Progress*, February 1992.

Boren, J. H. 1972. *When in Doubt, Mumble*, Van Nostrand Reinhold, New York.

Braverman, D. 1981. Fundamentals of Statistical Quality Control. Reston Publishing Co., Weston, VA.

Computer Programs. 1993. *Quality*, 32(6):48.

Computer Programs. 1994. *Quality*, 33(3):44.

Computer Programs. 1994. *Quality Progress*, 27(3):48.

Computer Programs. 1998. *Quality Progress*, 31(4):31.

Crosby, P. 1979. *Quality Is Free*, McGraw-Hill, New York.

Cruess, W. V. 1938. *Commercial Fruit and Vegetable Products*, McGraw-Hill, New York.

Daniels, S. 1998. *Quality Software Directory*, 31(4):27.

Deming, W. E. 1981. Unpublished lecture, San Diego, CA.

Deming, W. E. 1986. *Out of the Crisis*, Massachusetts Institute of Technology, Cambridge, MA.

Duncan, A. J. 1965. *Quality Control and Industrial Statistics*, Richard D. Irwin, Homewood, IL.

Ehrbar, A. 1993. Price of Progress. *Wall Street Journal*. 3/16/93.

Feigenbaum, A. V. 1983. *Total Quality Control, Engineering and Management*, McGraw-Hill, New York.

Florman, S. C. 1994. The human engineer. *Technology Review*, 97(7).

Golomski, W.A. 1993.Total quality management. *Food Technology*, 47(5):74–79.

Harmon, P., King, D. 1985. *Expert Systems*, John Wiley & Sons, New York.

Hollingsworth, P. 1997. Managing teams in the food industry. *Food Technology*, 51(11):75.

ISO 9000 Series. International Organization for Standardization.

Jackson, D. P. TQM does work! *Management Review*, May 1993.

Judge, E. E. *The Almanac,* Edward E. Judge and Sons, Westminster, MD.

Juran, J. M., Gryna, F. M. 1970. *Quality planning and analysis,* McGraw-Hill, New York.

Kane, V. E. 1989. *Defect Prevention,* Marcel Dekker, New York, New York.

Levy, S. Dr. Edelman's Brain, *New Yorker Magazine,* 5/2/94, p. 64.

Malcolm Baldrige National Quality Award. U.S. Dept. of Commerce, National Institute of Standards and Technology, Gaithersburg, MD.

Meyers, R. Quality Control Program Study for Community Colleges. California Community Colleges, Vocational Education Division, Sacramento, CA.

NBS Handbook 133. Checking the Net Contents of Packaged Goods. National Bureau of Standards, U.S. Department of Commerce. Washington, DC.

Peters, T., Waterman, Jr. 1982. *In Search of Excellence,* Harper and Row, New York.

Pierson, M. D., Corlett, D. A. 1992. *HACCP Principles and Applications,* Chapman and Hall, New York.

Quality Progress. 1990. The Tools of Quality. June through December.

Reed, S. F. *The Toxic Executive.* Harper Business.

Sayles, L. R. Why isn't TQM working? *Boardroom Reports,* March 15.

*SF Chronicle* 8/19/94; *SM Times* 8/18/94 Faulty Management Systems.

U.S. Department of Agriculture. Regulations Governing Inspection and Certification of Processed Fruits and Vegetables. Food Safety and Quality Service, Washington, DC.

U.S. Department of Agriculture. Code of Federal Regulations 9. Washington, DC.

U. S. Food and Drug Administration. *Code of Federal Regulations 21,* Washington DC.

Wagner, J. The myth and reality of TQM, *Frozen Food Report,* March/April 1993.

*Wall Street Journal,* August 8, 1994, Vol. CXXXI No. 26, Page 1.

MERTON HUBBARD is a consultant in food quality systems and failure analyses. His education includes degrees from M. I. T. and Stevens Institute of Technology, followed by 30 years of industrial experience with research, engineering, product safety, and quality control systems for such companies as Standard Brands, Ralston Purina, American Testing Institute, Hills Bros. Coffee (Nestlē), and dozens of clients in food processing, insurance, and law.

He holds registrations as Professional Engineer in the state of California in the fields of industrial engineering and quality engineering. Additionally, he has been elected Fellow in the American Society for Quality and carries their national certification as Quality Engineer.

He has been a frequent lecturer at national and international conferences and is an instructor in quality control at various colleges and industries.